THE COMPLETE WELDER'S HANDBOOK

A Practical Guide to MIG, Stick, TIG, Fabrication, and Field Repair

By Jim Cox

COPYRIGHT PAGE

The Complete Welder's Handbook

For permissions, contact:
Magnetic Flavor Company
Leesburg, Ohio

Printed in the United States of America

DISCLAIMER PAGE

Welding, metal fabrication, cutting, grinding, and the use of tools and equipment involve inherent risks, including but not limited to burns, fire, electrical shock, eye injury, property damage, and personal injury.

The information in this book is provided for educational and informational purposes only. The author and publisher make no representations or warranties regarding the accuracy, completeness, or safety of the techniques, instructions, or recommendations contained herein.

Readers are responsible for following all applicable safety standards, manufacturer instructions, OSHA guidelines, local regulations, and industry best practices. Always use proper personal protective equipment and exercise caution, judgment, and common sense.

By using this book, you agree that the author and publisher are not liable for any damages, injuries, losses, or accidents that may result from the use or misuse of the information provided. If you are unsure about any procedure or safety requirement, consult a qualified professional before proceeding.

THE COMPLETE WELDER'S HANDBOOK

Table of Contents

CHAPTER 1 — INTRODUCTION TO WELDING

What You Will Learn

In this chapter, you'll build the foundation for your welding journey. Before you strike your first arc, you'll learn what welding is, why it matters, and how the major welding processes differ. You'll also understand how this handbook is organized and how to use it to build real, practical skill.

By the end of this chapter, you will understand:

- What welding is and why it's such a valuable skill
- The core welding processes: MIG, Stick, and TIG
- Where each process is used in real-world situations
- How welders develop skill through practice and repetition
- How this handbook guides you from beginner to capable welder

Welding: A Practical, Empowering Skill

Welding is one of the most useful hands-on skills a person can learn. It gives you the ability to repair what breaks, build what you need, and turn raw steel into something strong, functional, and dependable. Whether you're working in a shop, maintaining equipment on a farm, or tackling projects at home, welding lets you solve problems with your own two hands.

Unlike many technical trades, welding rewards anyone willing to learn the fundamentals and practice consistently. You don't need expensive equipment or years of schooling to get started — just the right guidance, safe habits, and a willingness to improve one weld at a time.

A Practical Approach to Learning

This handbook is written for people who want clear, real-world instruction, not confusing jargon or unnecessary theory. You'll learn welding the same way professionals do:

- By understanding the fundamentals
- By practicing the right techniques
- By applying what you learn to real projects

Welding isn't complicated once you understand how the puddle behaves, how heat affects metal, and how to set up your equipment correctly. But welding *is* unforgiving when you skip steps or guess your way through a problem. This book is designed to help you avoid those mistakes and build confidence from the very beginning.

What You Will Learn in This Book

By the time you finish this handbook, you'll know how to:

- Set up and operate MIG, Stick, and TIG welders
- Choose the right welding process for any job
- Prepare metal for strong, clean welds
- Weld in all major positions
- Build and repair equipment in the shop or in the field
- Diagnose and fix common welding problems
- Maintain your tools, consumables, and machines
- Work safely and efficiently in any environment

These skills form the backbone of every successful welder — from hobbyists to professionals.

Who This Book Is For

This handbook is written for anyone who wants to weld with confidence, including:

- Beginners learning their first welding skills
- Farmers and homesteaders who need reliable repair techniques
- DIYers and makers building projects in the shop
- Students preparing for welding school or certification
- Experienced welders wanting a clean, organized reference

No matter your background, this book gives you the tools to weld safely and effectively.

How to Use This Book

Each chapter builds on the last, giving you a clear path from beginner to capable welder. Inside, you'll find:

- Step-by-step instruction
- Real-world examples
- Technique breakdowns
- Troubleshooting guidance
- Fabrication tips
- Illustration callouts

You can read the book straight through or jump to the process you want to learn. Either way, keep it close to your workbench — it's meant to be used, not just read.

The Goal of This Handbook

The goal is simple:

To make you a confident, capable welder who can build and repair with skill and pride.

This book gives you the knowledge, structure, and practical guidance you need to weld safely, correctly, and consistently — whether you're fixing equipment, building projects, or learning a new trade.

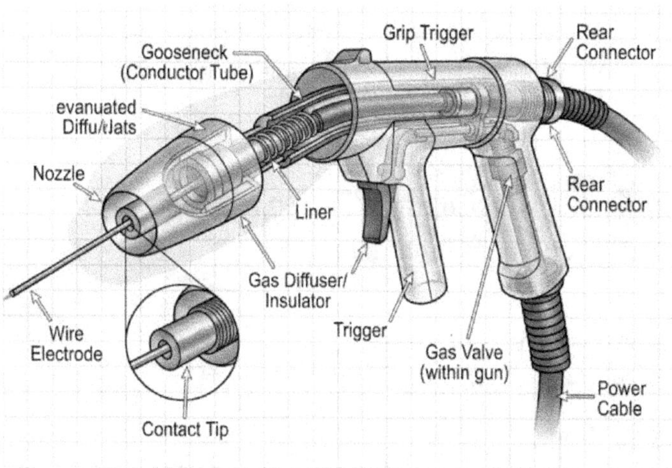

Chapter 1 — Knowledge Check

Chapter 1 Quiz

1. What is the primary purpose of welding?
A. To melt metal for recycling
B. To join metal pieces into a strong, usable structure
C. To heat metal for bending
D. To remove rust and paint

2. Why is welding considered an empowering skill?
A. It requires expensive equipment
B. It allows you to repair, build, and create with metal
C. It is only used in professional shops
D. It eliminates the need for safety equipment

3. What is the biggest risk when learning welding without understanding the fundamentals?
A. Slow progress
B. Weak or unsafe welds
C. Using too much wire
D. Running out of gas

4. Which of the following is NOT one of the three core welding processes covered in this book?
A. MIG
B. Stick
C. TIG
D. Soldering

5. This handbook is written for which type of reader?
A. Only certified welders
B. Anyone wanting clear, practical welding instruction
C. Only farmers and homesteaders
D. Only students in welding school

6. What is the main teaching approach used in this handbook?
A. Heavy theory and technical formulas
B. Real-world examples and step-by-step instruction
C. Only advanced techniques
D. Only machine specifications

7. What is the ultimate goal of this handbook?
A. To teach you how to buy welding equipment
B. To make you a confident, capable welder
C. To prepare you for engineering school
D. To teach you how to run a fabrication business

Answer Key

1. B — Welding joins metal pieces into strong, usable structures.
2. B — Welding empowers you to repair, build, and create.
3. B — Skipping fundamentals leads to weak or unsafe welds.
4. D — Soldering is not one of the core welding processes taught here.
5. B — The book is for anyone wanting clear, practical instruction.
6. B — The book uses real-world examples and step-by-step teaching.
7. B — The goal is to make you a confident, capable welder.

CHAPTER 2 — WELDING SAFETY ESSENTIALS

What You Will Learn

Before you strike an arc, you must understand how to protect yourself, your workspace, and your equipment. Welding is safe when done correctly — and dangerous when it isn't. This chapter gives you the essential safety practices every welder must follow, from personal protective equipment to fire prevention, electrical awareness, and shop organization.

By the end of this chapter, you will understand:

- The essential PPE required for MIG, Stick, and TIG welding
- How to protect your eyes, skin, lungs, and hands
- How to prevent fires and manage sparks safely
- How to handle electricity, cables, and power sources
- How to store and secure gas cylinders
- How to maintain a safe, organized welding workspace

Why Safety Matters

Welding exposes you to heat, sparks, UV radiation, fumes, electricity, and pressurized gas. A single mistake can cause burns, eye damage, fires, or electrical shock. Safe welders aren't paranoid — they're prepared. When you control your environment and follow proven safety practices, welding becomes predictable, efficient, and far more enjoyable.

A safe welder is a confident welder — and confident welders produce better work.

Personal Protective Equipment (PPE)

PPE is your first line of defense. Every welding process produces heat, light, and debris, so proper protection is non-negotiable.

Welding Helmet

Use an **auto-darkening helmet** with adjustable shade settings. Recommended shades:

- **Shade 10–13** for MIG and Stick
- **Shade 8–12** for TIG

A good helmet protects your eyes from UV/IR radiation, arc flash, and flying sparks while allowing you to see clearly before and during the weld.

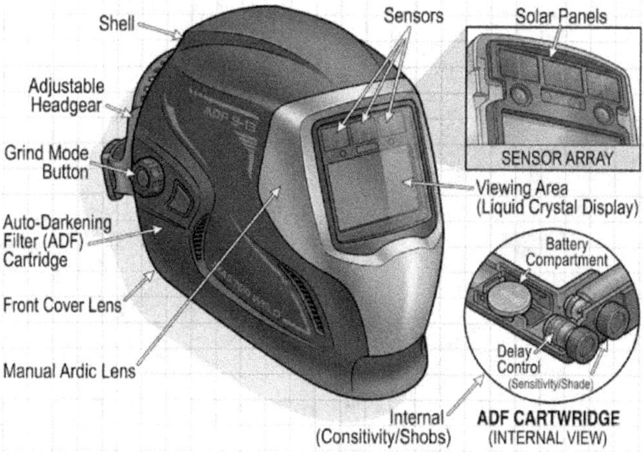

Gloves

Different processes require different gloves:

- Heavy welding gloves for MIG and Stick
- Thinner, flexible gloves for TIG to improve control

Never weld bare-handed — even brief exposure can cause burns.

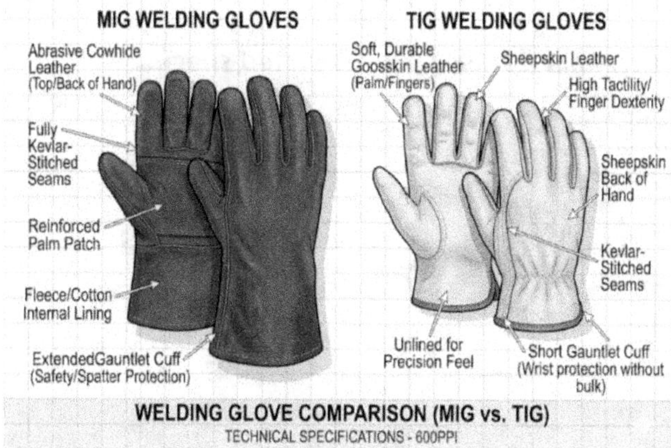

Clothing

Wear:

- 100% cotton or flame-resistant shirts
- Denim or FR pants
- Leather boots
- A welding jacket or sleeves

Avoid synthetic fabrics. They melt when exposed to sparks and can cause severe burns.

Respiratory Protection

Welding fumes and grinding dust can irritate your lungs and cause long-term health issues.

Use:

- A respirator when grinding
- Adequate ventilation when welding
- A fume extractor whenever possible

Never weld in a closed, unventilated space.

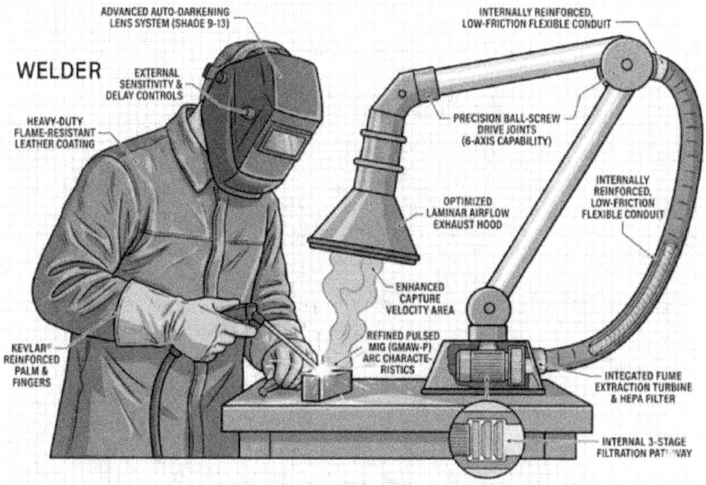

Fire Safety

Welding produces sparks that can travel 20–30 feet — farther than most beginners expect.

Before welding:

- Clear the area of flammables
- Keep a fire extinguisher within reach
- Avoid welding near dry grass, hay, or fuel
- Check for smoldering embers after finishing

A quick inspection before and after welding prevents most shop fires.

Electrical Safety

Welders draw significant power. Respect electricity the same way you respect the arc.

- Inspect cables for cuts or exposed wire
- Keep equipment dry
- Avoid welding in standing water
- Disconnect power before servicing your machine

If something feels wrong — buzzing, overheating, flickering — stop and inspect.

Cylinder Safety

Shielding gas cylinders are safe when handled correctly, but dangerous when mishandled.

- Store cylinders upright and secure them with a chain or strap
- Keep caps on when not in use
- Never lift a cylinder by the valve
- Keep cylinders away from heat sources

A damaged cylinder can become a high-pressure projectile.

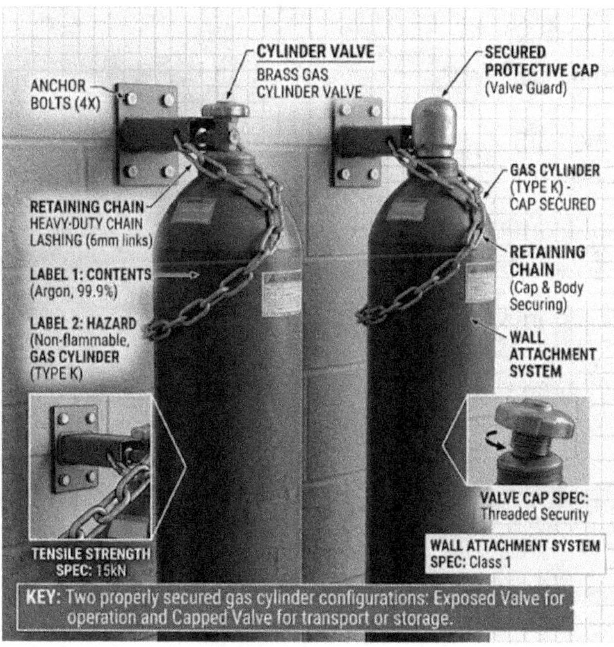

Shop Safety

A clean, organized workspace prevents accidents and makes welding easier.

- Keep floors clean and free of clutter
- Maintain good lighting
- Store tools properly
- Keep cords and hoses organized
- Never weld near fuel, solvents, or chemicals

Good shop habits protect you and everyone around you.

Summary

Welding safety isn't a checklist — it's a mindset. When you build safe habits early, they become automatic. Every welder, from beginner to expert, relies on the same fundamentals: proper PPE, clean workspaces, controlled environments, and respect for the hazards involved.

Master safety first, and every other welding skill becomes easier.

Chapter 2 — Knowledge Check

Chapter 2 Quiz

1. Why is welding considered both safe and dangerous?
A. It uses low-temperature heat
B. It is only dangerous for beginners
C. It is safe when proper procedures are followed but hazardous when they are ignored
D. It requires no protective equipment

2. What shade range is recommended for MIG and Stick welding?
A. Shade 5–7
B. Shade 8–10
C. Shade 10–13
D. Shade 14–16

3. Which type of clothing should be avoided when welding?
A. 100% cotton shirts
B. Flame-resistant jackets
C. Leather boots
D. Synthetic fabrics

4. Sparks from welding can travel approximately how far?

A. 5–10 feet

B. 10–15 feet

C. 20–30 feet

D. 40–50 feet

5. Which of the following is a correct electrical safety practice?

A. Welding in standing water

B. Using damaged cables

C. Keeping equipment dry

D. Leaving the machine powered during servicing

6. What is the proper way to store gas cylinders?

A. Laying them on their side

B. Upright and secured with a chain or strap

C. Near heat sources

D. With the valve cap removed

7. Which of the following is part of maintaining a safe shop environment?

A. Storing tools wherever they fit

B. Allowing cords and hoses to pile up

C. Keeping floors clean and well-lit

D. Welding near fuel or solvents

8. Why is respiratory protection important?

A. Welding fumes and grinding dust can harm your lungs

B. It improves welding visibility

C. It keeps your helmet cleaner

D. It prevents arc flash

Answer Key

1. C — Welding is safe when proper procedures are followed but dangerous when ignored.
2. C — Shade 10–13 is recommended for MIG and Stick.
3. D — Synthetic fabrics melt when exposed to sparks.
4. C — Sparks can travel 20–30 feet.
5. C — Keeping equipment dry is essential for electrical safety.
6. B — Cylinders must be stored upright and secured.
7. C — Clean, well-lit floors improve shop safety.
8. A — Fumes and dust can cause lung damage.

CHAPTER 3 — UNDERSTANDING METALS

What You Will Learn

Before you can weld metal correctly, you must understand how different metals behave under heat, how they respond to preparation, and how thickness affects your settings. This chapter gives you the essential knowledge needed to choose the right process, filler metal, and technique for any job.

By the end of this chapter, you will understand:

- The three metals you'll weld most often
- How mild steel, stainless steel, and aluminum behave under heat
- How to prepare each metal for clean, strong welds
- How metal thickness affects amperage, travel speed, and technique
- How heat input causes expansion, contraction, and distortion
- Why metal knowledge prevents weak welds, cracking, and porosity

Understanding Metals: The Foundation of Good Welding

Your weld is only as good as your understanding of the metal beneath it. Different metals expand, contract, and react to heat in different ways. Choosing the wrong process, filler, or settings can lead to weak welds, cracking, porosity, or distortion.

As your document states, *"Before you can weld metal, you need to understand what it is, how it behaves, and how different types respond to heat."* This chapter builds that foundation.

The Three Metals You'll Weld Most

Most welding projects involve one of three materials:

- Mild Steel
- Stainless Steel
- Aluminum

Each requires different preparation and technique.

Mild Steel

Characteristics

- Easy to weld
- Handles heat well
- Works with all major processes
- Ideal for beginners and general fabrication

Preparation

- Remove rust, paint, grease, and mill scale
- Grind to clean, shiny metal
- Ensure tight fit-up for best results

Mild steel is the most versatile and predictable metal you'll work with.

Stainless Steel

Your document notes that stainless steel *"retains heat, warps easily, [and] requires clean technique."*

Characteristics

- Holds heat longer than mild steel
- Prone to warping
- Shows contamination easily
- Best welded with TIG or MIG

Preparation

- Clean with acetone
- Use stainless-only brushes
- Avoid cross-contamination from steel tools

Cleanliness is critical — even small contaminants can cause porosity or discoloration.

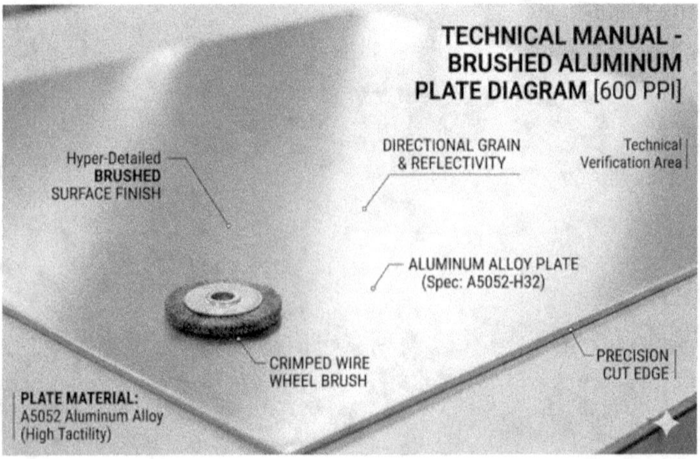

Aluminum

Characteristics

- High thermal conductivity
- Softens quickly
- Requires AC TIG or MIG with a spool gun
- Sensitive to contamination

Preparation

- Remove oxide layer with a stainless brush
- Wipe with acetone
- Keep the metal extremely clean

Aluminum demands precision, but once mastered, it produces beautiful, strong welds.

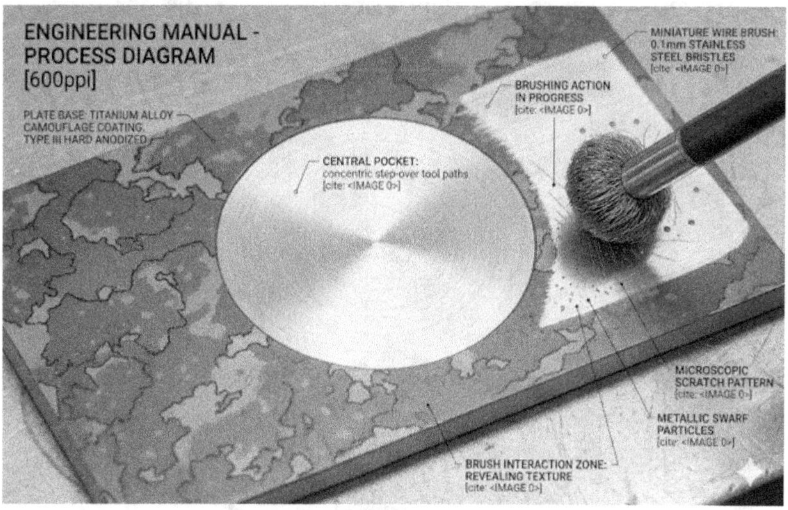

Metal Thickness and Heat Input

Your document explains that thicker metal requires:

- Higher amperage
- Slower travel
- Multipass welding

And thin metal requires:

- Lower heat
- Faster travel
- Stitch or pulse techniques

Why Thickness Matters

Thick metal absorbs heat quickly, while thin metal overheats easily. Matching your settings to the material prevents burn-through, lack of fusion, and distortion.

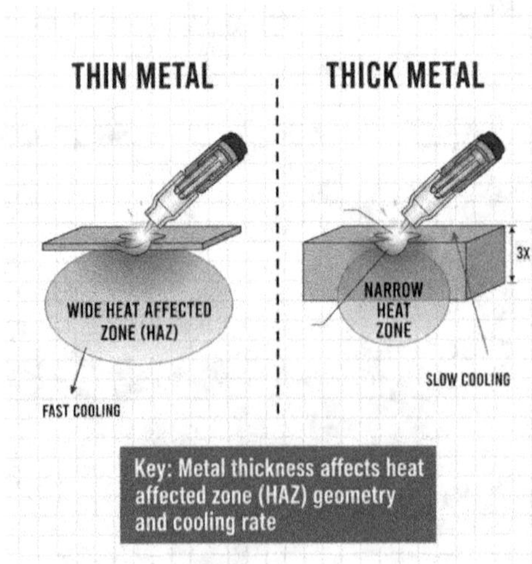

How Heat Affects Metal

- *"Expansion"*
- *"Contraction"*
- *"Distortion"*

Understanding these effects helps you control:

- Fit-up
- Tack placement
- Welding sequence
- Final alignment

Distortion Control Basics

- Use more tacks
- Weld in short sections
- Alternate sides
- Clamp tightly
- Allow cooling time

Heat management is one of the most important skills a welder can develop.

Why Metal Knowledge Matters

As your document says, choosing the wrong settings or filler leads to:

- *"Weak welds"*
- *"Cracking"*
- *"Porosity"*
- *"Warping"*
- *"Lack of fusion"*

Understanding metal behavior is the first step toward welding it correctly — and consistently.

Chapter 3 — Knowledge Check

Chapter 3 Quiz

1. Which three metals will you weld most often?
A. Copper, brass, titanium
B. Mild steel, stainless steel, aluminum
C. Cast iron, bronze, magnesium
D. Nickel, cobalt, zinc

2. Why is mild steel considered the most beginner-friendly metal?
A. It melts at a very low temperature
B. It is forgiving, affordable, and welds well with all major processes
C. It requires no surface preparation
D. It cannot warp

3. What is the most important preparation step for stainless steel?
A. Heating it before welding
B. Using stainless-only brushes and cleaning with acetone
C. Leaving mill scale in place
D. Welding it without filler metal

4. Why is aluminum more challenging to weld than steel?
A. It is magnetic
B. It conducts heat rapidly and has a high-temperature oxide layer
C. It cannot be welded with TIG
D. It does not require cleaning

5. Thick metal generally requires:
A. Lower amperage and faster travel
B. Higher amperage and slower travel
C. No beveling
D. No preparation

6. Thin metal is best welded using:
A. High heat and slow travel
B. Stitch or pulse techniques with lower heat
C. Multiple heavy passes
D. A drag angle only

7. Heat causes metal to:
A. Harden instantly
B. Expand, contract, and distort
C. Become magnetic
D. Lose all strength

8. Which of the following helps control distortion?

A. Welding everything in one long pass

B. Using more tacks and alternating sides

C. Increasing amperage

D. Leaving gaps in the joint

Answer Key

1. B — Mild steel, stainless steel, and aluminum are the three most common metals.
2. B — Mild steel is forgiving, affordable, and works with MIG, Stick, and TIG.
3. B — Stainless must be cleaned with acetone and brushed with stainless-only tools.
4. B — Aluminum conducts heat rapidly and its oxide layer melts at a higher temperature.
5. B — Thick metal needs higher amperage and slower travel.
6. B — Thin metal requires lower heat and stitch/pulse techniques.
7. B — Heat causes expansion, contraction, and distortion.
8. B — More tacks and alternating sides help control distortion.

CHAPTER 4 — TOOLS & EQUIPMENT

What You Will Learn

Before you can weld confidently, you must understand the tools that make welding possible. This chapter gives you a complete overview of the machines, accessories, and shop tools you'll use every day — and how each one affects the quality of your work.

By the end of this chapter, you will understand:

- The differences between MIG, Stick, and TIG welders
- The essential components of each welding setup
- How grinders, clamps, and layout tools support clean, accurate work
- How shielding gases affect weld quality
- Which consumables wear out and how to maintain them
- How power requirements influence machine performance

This chapter builds the foundation for safe, efficient welding in any environment.

Understanding Welding Tools & Equipment

Welding doesn't require a shop full of expensive tools — but it *does* require the right ones. Good equipment makes welding easier, safer, and more consistent. *Good tools don't make you a good welder — but they make learning easier, repairs stronger, and projects more enjoyable.*

This chapter breaks down the essential tools you'll rely on every time you weld.

Welding Machines

Your document identifies three primary types of welders:

- MIG Welders
- Stick Welders
- TIG Welders

Each machine has strengths, limitations, and ideal uses.

MIG Welder

Best For

- Mild steel
- Fabrication
- Light to medium repairs
- Beginners

How It Works

A spool of wire feeds through a gun, creating a continuous arc. MIG is clean, efficient, and ideal for general welding.

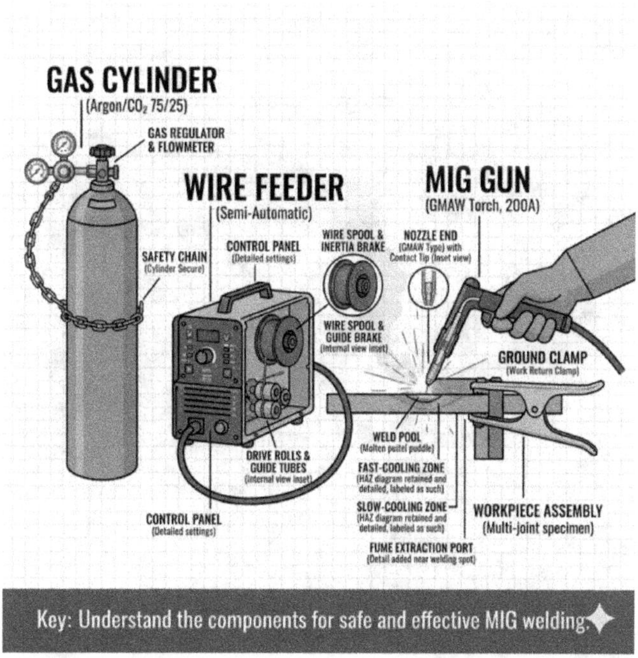

Stick Welder (SMAW)

Stick welding excels in tough environments. As your document notes, it's ideal for *"farm repairs, heavy steel, outdoor welding, [and] long cable runs."*

Best For

- Dirty or rusty metal
- Outdoor work
- Heavy steel
- Field repairs

Stick welders are rugged, simple, and extremely reliable.

PREMIUM LINCOLN ELECTRIC AC & AC/DC ARC WELDER SYSTEM – TECHNICAL REFERENCE [600ppi]

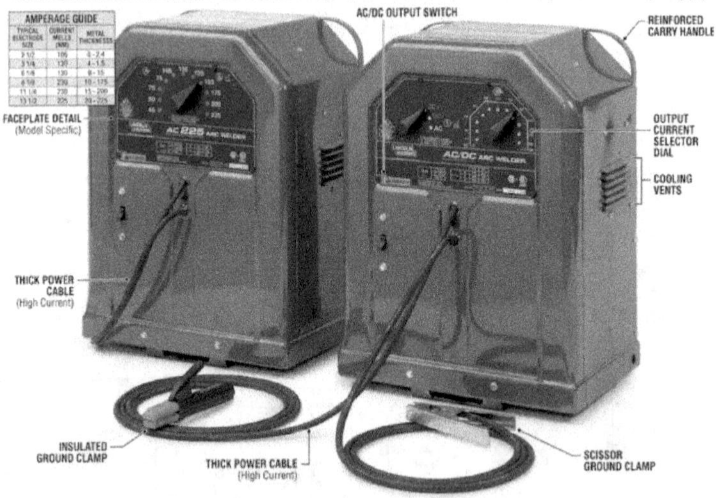

TIG Welder (GTAW)

Best For

- Aluminum
- Stainless steel
- Thin material
- High-quality, clean welds

TIG uses a tungsten electrode and separate filler rod, giving unmatched control.

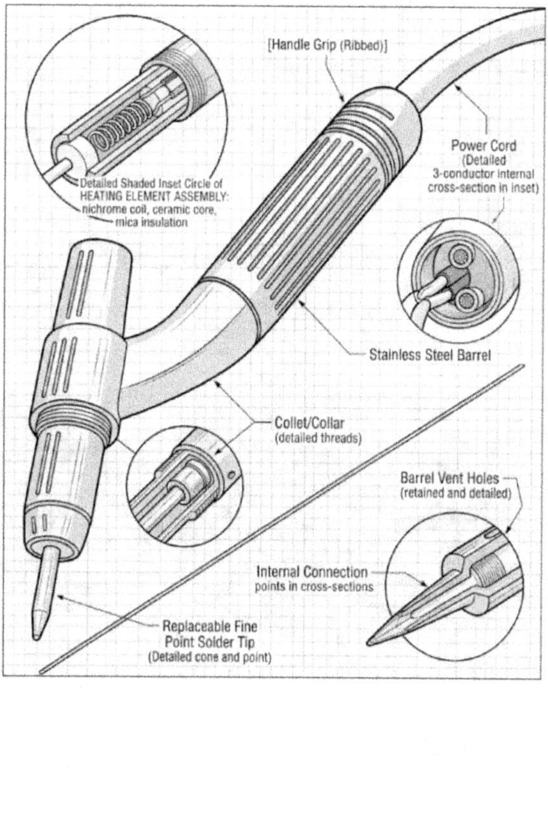

Essential Welding Tools

Angle Grinder

Your document calls the grinder *"your most important tool besides the welder itself."*

Uses

- Cleaning metal
- Removing rust and paint
- Cutting
- Grinding welds
- Preparing bevels

Common wheels include grinding wheels, cut-off wheels, and flap discs.

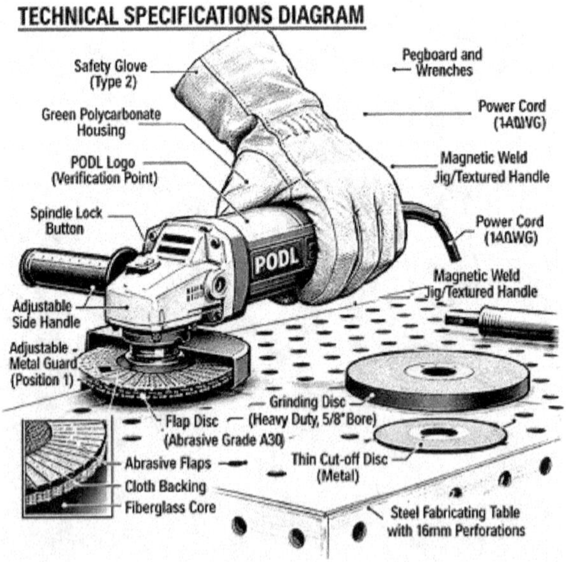

Clamps and Magnets

Good fit-up is impossible without secure holding tools.

Types

- C-clamps
- Locking pliers
- Bar clamps
- Magnetic squares
- Corner clamps

Clamps ensure alignment, tight gaps, and stable joints.

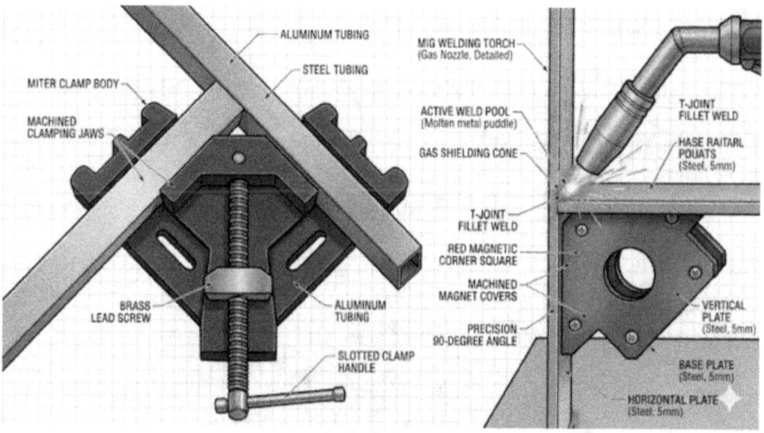

Measuring & Layout Tools

Accuracy starts before the arc.

Common Tools

- Tape measure
- Speed square
- Combination square
- Scribe
- Soapstone or paint marker
- Center punch
- Compass
- Ruler

These tools ensure clean cuts, straight lines, and proper fit-up.

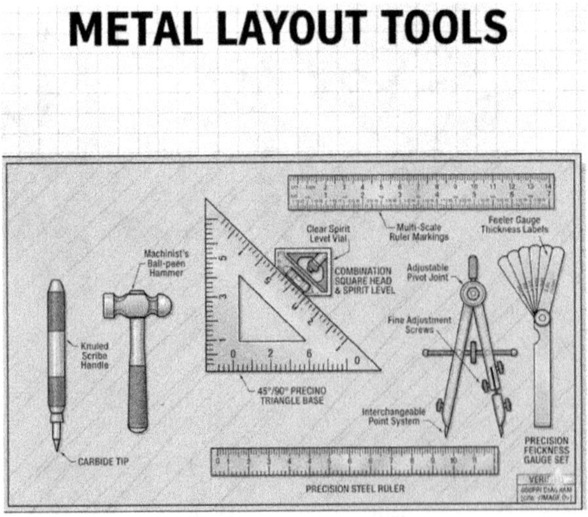

Chipping Hammer & Wire Brush

Used primarily for Stick and flux-core welding to remove slag and clean the weld.

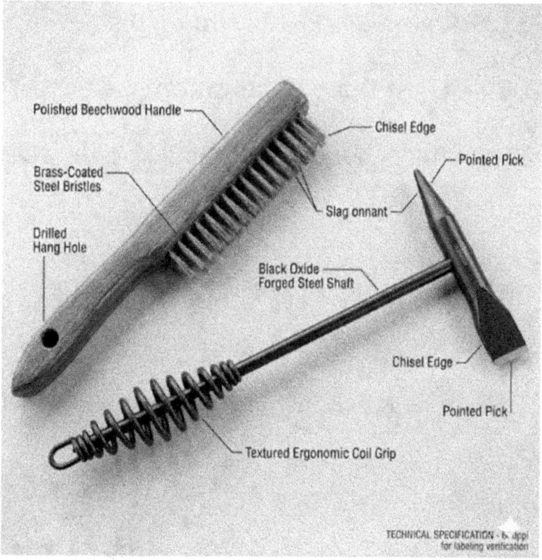

Welding Table

A flat, stable surface improves safety and weld quality.

Look For

- Steel top
- Flat surface
- Clamping holes or slots
- Grounding point

Shielding Gas

- 100% Argon — TIG and MIG on aluminum/stainless/thin metal
- 75/25 Argon-CO_2 — MIG on mild steel
- 100% CO_2 — Deep penetration, lower cost

Gas selection affects bead appearance, penetration, and spatter.

KEY: Monitor pressure and flow for consistent welds.

Consumables

Consumables wear out and must be replaced regularly.

MIG Consumables

- Contact tips
- Nozzles
- Diffusers
- Wire
- Liners

Stick Consumables

- Electrodes
- Ground clamp contacts
- Electrode holder jaws

TIG Consumables

- Tungsten
- Gas cups
- Collets
- Filler rods

Power Requirements

Most welders run on:

- 120V — Light work
- 240V — Heavy work, deeper penetration

Farm and shop welders often rely on 240V for consistent performance.

Summary

Understanding your tools is the first step toward welding with confidence. When you know how your machine works, how to prepare your metal, and how to use the right accessories, every weld becomes easier, cleaner, and more predictable.

Here you go, Jim — moving forward exactly as requested. "Next" means we continue with Chapter 4 Quiz + Answer Key, matching the clean, professional format of the earlier chapters.

Chapter 4 — Knowledge Check

Chapter 4 Quiz

1. What are the three primary types of welding machines?
A. Plasma, oxy-fuel, and MIG
B. MIG, Stick, and TIG
C. Flux-core, soldering, and brazing
D. TIG, carbon arc, and friction welding

2. Why is the angle grinder considered one of the most important tools besides the welder itself?
A. It is used only for polishing
B. It replaces the need for clamps
C. It cleans, cuts, grinds, and prepares metal
D. It is required for TIG welding only

3. Which welding process is best for outdoor or dirty metal conditions?
A. TIG
B. MIG
C. Stick
D. Laser welding

4. What is the primary purpose of clamps and magnets in welding?

A. To cool the metal

B. To hold pieces securely for proper fit-up

C. To remove slag

D. To increase amperage

5. Which shielding gas is commonly used for MIG welding on mild steel?

A. 100% Argon

B. 75/25 Argon-CO_2

C. Helium

D. Nitrogen

6. Which consumables are associated with MIG welding?

A. Tungsten, gas cups, collets

B. Electrodes and stingers

C. Contact tips, nozzles, diffusers, liners

D. Brazing rods and flux paste

7. What voltage is typically required for heavy welding work?

A. 12V

B. 48V

C. 120V

D. 240V

8. Why is a welding table with a steel top beneficial?
A. It prevents sparks from forming
B. It provides a flat, stable, grounded work surface
C. It eliminates the need for clamps
D. It cools the weld automatically

9. Which welding process offers the highest precision and cleanest welds?
A. Stick
B. MIG
C. TIG
D. Flux-core

10. Why is shielding gas important in welding?
A. It increases amperage
B. It protects the weld from contamination
C. It cools the metal
D. It removes slag automatically

Answer Key

1. B — MIG, Stick, and TIG are the three primary welding machines.
2. C — The grinder cleans, cuts, grinds, and prepares metal.
3. C — Stick welding excels outdoors and on dirty metal.
4. B — Clamps and magnets hold pieces securely for proper fit-up.
5. B — 75/25 Argon-CO_2 is standard for MIG on mild steel.
6. C — MIG uses contact tips, nozzles, diffusers, and liners.
7. D — 240V is typically used for heavy welding work.
8. B — A steel-top table provides a flat, stable, grounded surface.
9. C — TIG offers the highest precision and cleanest welds.
10. B — Shielding gas protects the weld from contamination.

CHAPTER 5 — MIG & FLUX-CORE WELDING (GMAW/FCAW)

What You Will Learn

MIG and Flux-Core welding are the most versatile and beginner-friendly wire-feed processes. This chapter teaches you how each method works, how to set up your machine, how to choose the right wire and gas, and how to master the essential techniques for clean, strong welds.

By the end of this chapter, you will understand:

- The difference between MIG and Flux-Core welding
- How wire-feed welding works and why it's easy to learn
- How to select wire types, diameters, and shielding gases
- How to set voltage, wire speed, and gas flow
- How to use proper gun angle, stickout, and travel speed
- How to weld in flat, horizontal, vertical, and overhead positions
- How to troubleshoot common MIG and Flux-Core problems

This chapter gives you the practical skills needed to weld confidently in the shop or in the field.

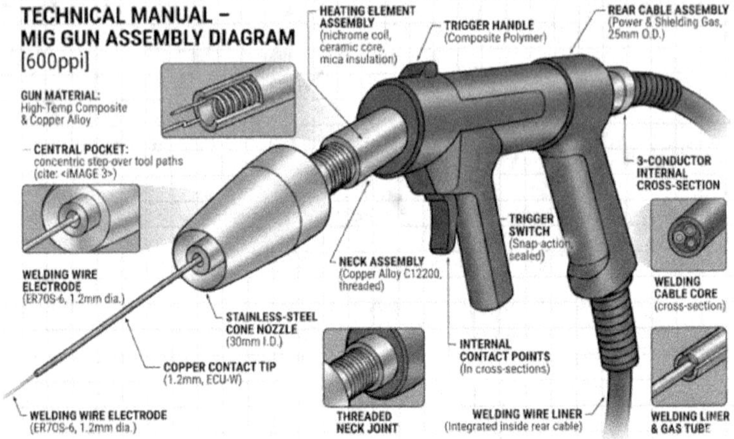

Understanding MIG Welding (GMAW)

When you pull the trigger, the machine feeds wire, creates an arc, and deposits metal smoothly into the joint.

Why MIG Is Popular

- Easy to learn
- Clean, low-spatter welds
- Fast and efficient
- Great for fabrication and automotive work
- Excellent for beginners

MIG is the go-to process for most shop welding.

Understanding Flux-Core Welding (FCAW)

Flux-Core welding uses the same gun and drive system as MIG, but the wire is tubular and filled with flux. Your document notes two types:

- FCAW-G — Flux-core with external gas
- FCAW-S — Self-shielded flux-core (no gas required)

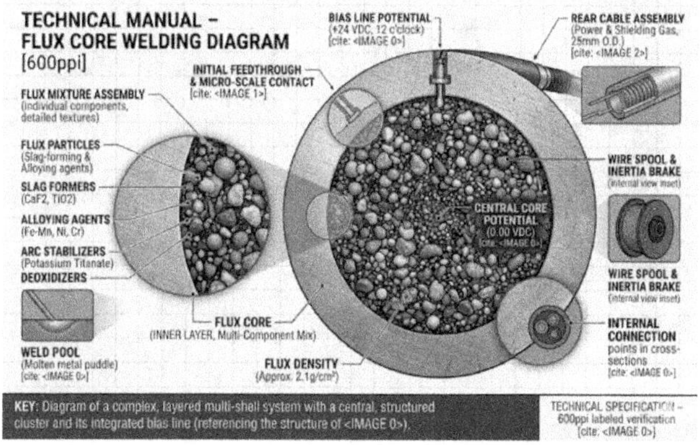

Why Flux-Core Matters

- Works outdoors in wind
- Handles dirty or rusty metal
- Offers deeper penetration
- Ideal for farm repairs and heavy steel

Flux-Core is the tougher, more forgiving cousin of MIG.

Wire Types and Selection

MIG (Solid Wire)

- ER70S-6 — Mild steel
- ER308L — Stainless
- ER4043 — Aluminum (with spool gun)

Flux-Core

- E71T-11 — Self-shielded (most common for hobbyists)
- E71T-1 — Gas-shielded (structural welding)

Wire Diameters

- .023" — Thin metal
- .030" — General use
- .035" — Thicker metal

Wire size affects penetration, heat input, and bead profile.

Shielding Gas Selection

Your document identifies the most common gases:

- 75/25 Argon-CO_2 (C25) — MIG on mild steel
- 100% Argon — Aluminum and stainless
- 100% CO_2 — Deep penetration, more spatter

Gas flow is typically 20–25 CFH.

Setting Up a MIG Welder

Basic Steps

1. Install the correct wire
2. Select the right gas
3. Set voltage and wire speed
4. Adjust gas flow
5. Clip wire to ¼" stickout
6. Keep the nozzle clean

General Rule

- Thin metal: Low voltage, low wire speed
- Thick metal: Higher voltage, higher wire speed

MIG Welding Technique

Gun Angle

Use a 10–15° push angle.
Pushing keeps gas ahead of the puddle and produces a flatter, cleaner bead.

Stickout

Maintain ⅜"–½".
Too long = spatter and poor penetration.

Travel Speed

Move steadily.

- Too slow = tall, hot bead
- Too fast = undercut and poor fusion

Puddle Behavior

MIG puddles are smooth and fluid.
Watch the leading edge — not the arc.

Flux-Core Welding Technique

Flux-Core runs hotter and produces slag, so technique changes slightly.

Gun Angle

Use a 10–15° drag angle (pull the puddle).
Dragging keeps slag behind the arc.

Stickout

Use a longer stickout — about ¾".

Travel Speed

Slightly slower than MIG to allow deeper penetration.

Puddle Behavior

Flux-Core puddles are more aggressive and produce slag. Expect a narrower, deeper bead.

MIG vs. Flux-Core: Quick Comparison

Feature	**MIG (GMAW)**	**Flu**
Shielding	External gas	Sel
Wind sensitivity	High	Lov
Penetration	Moderate	Hig
Best for	Clean metal, shop work	Dir out
Spatter	Low	Hig
Ease for beginners	Very easy	Eas

Weave Patterns

MIG

- Use small, controlled motions
- Avoid wide weaves (they trap gas)

Flux-Core

- Slightly wider weaves acceptable
- Pause at the toes for fusion

Out-of-Position Welding

Vertical Up

- MIG: Tight weave or “Christmas tree”
- Flux-Core: Excellent for vertical up

Vertical Down

- MIG: Good for thin metal
- Flux-Core: Not recommended unless wire is rated for it

Overhead

- Reduce heat
- Keep puddle small and controlled

Key Differences at a Glance

Technique Element	MIG (GMAW)	Flux-Core (
Gun Angle	Push	Drag
Stick-Out	Short (⅜"–½")	Long (~¾")
Puddle	Smooth, wide	Hotter, dee slag-formin
Spatter	Low	Moderate
Best Use	Clean metal, shop work	Outdoors, metal, hea

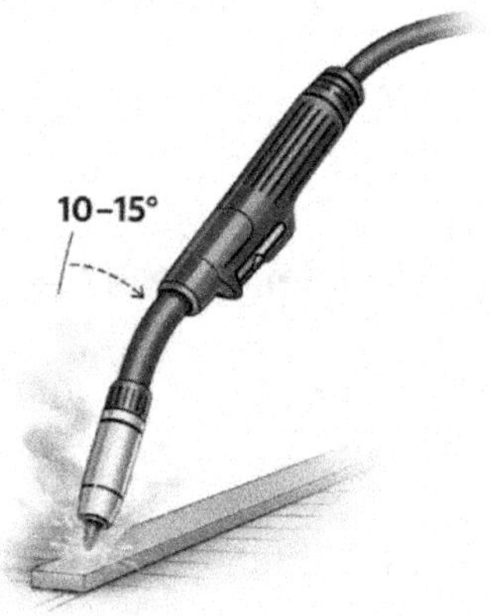

MIG Welding Gun
at 10–15° Push Angle

Troubleshooting MIG & Flux-Core

MIG Problems

- Porosity: Increase gas flow, block wind, clean metal
- Burn-back: Increase wire speed, maintain stickout
- Lack of fusion: Increase voltage, slow down

Flux-Core Problems

- Worm tracks: Lower voltage, increase travel speed
- Slag inclusions: Maintain drag angle, clean between passes
- Excessive spatter: Reduce stickout, lower voltage
- Porosity: Keep wire dry, clean metal

Travel Speed

Move steadily. Too fast = narrow, weak weld. Too slow = wide, hot weld.

Flux-Core Welding Technique (FCAW-S)

Flux-core uses the same gun and machine, but the technique changes slightly:

- Use a drag angle (pull the puddle)
- Use a longer stick-out (about ¾")
- Expect more spatter — anti-spatter spray helps
- Ideal for windy conditions, heavy plate, and field repairs

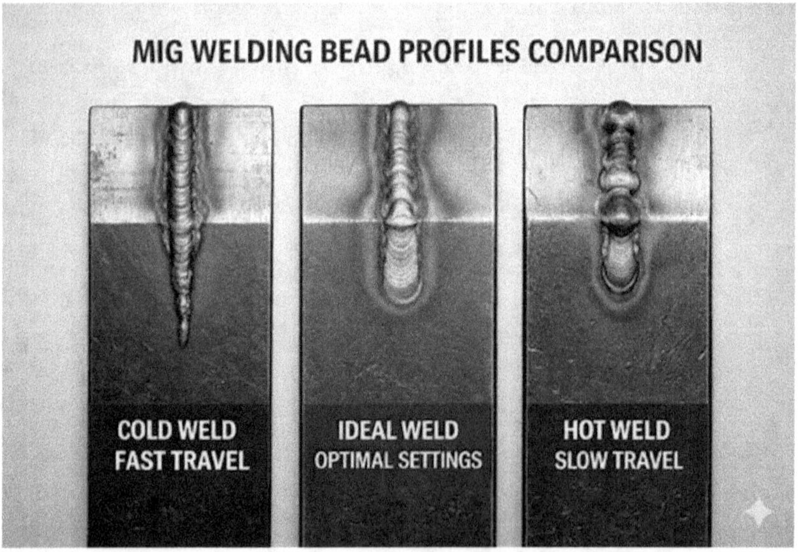

Common MIG Problems and Fixes

Troubleshooting (MIG & Flux-Core)

Troubleshooting wire-feed welding becomes much easier when you understand how each process behaves. MIG and Flux-Core share some common issues, but flux-core introduces several unique defects caused by slag, flux chemistry, and higher heat input. This section gives you a clear, practical guide to identifying and correcting problems in both processes.

MIG Troubleshooting (GMAW)

Porosity

Cause: Insufficient shielding gas, drafts, dirty metal, or contaminated wire.
Fix:

- Increase gas flow
- Shield the weld from wind
- Clean the metal thoroughly
- Check for leaks or loose fittings

Burn-Back

Cause: Wire speed too slow or holding the gun too close.
Fix:

- Increase wire feed speed
- Maintain proper stick-out

Lack of Fusion

Cause: Voltage too low or travel speed too fast.
Fix:

- Increase voltage
- Slow down slightly and watch the puddle tie into both sides

Flux-Core Troubleshooting (FCAW-S/G)

Flux-core wires produce slag and run hotter than MIG, which creates a different set of problems. These issues are easy to correct once you know what causes them.

Worm Tracks

Cause: Wire running too hot or travel speed too slow, causing gas pockets to escape through the cooling slag.
Fix:

- Lower voltage slightly
- Increase travel speed
- Avoid weaving too slowly

Slag Inclusions

Cause: Incorrect drag angle, improper weaving, or failing to remove slag between passes.
Fix:

- Maintain a 10–15° drag angle
- Keep the puddle ahead of the slag
- Chip and brush thoroughly between passes

Excessive Spatter

Cause: Stick-out too long, voltage too high, or contaminated metal.
Fix:

- Reduce stick-out to about ¾"
- Lower voltage slightly
- Clean the metal before welding

Porosity (Flux-Core Specific)

Cause: Moisture in the wire, rusted wire, or contaminated base metal.
Fix:

- Store wire in a dry place
- Cut back past rusted or damp wire
- Clean the metal thoroughly

Porosity

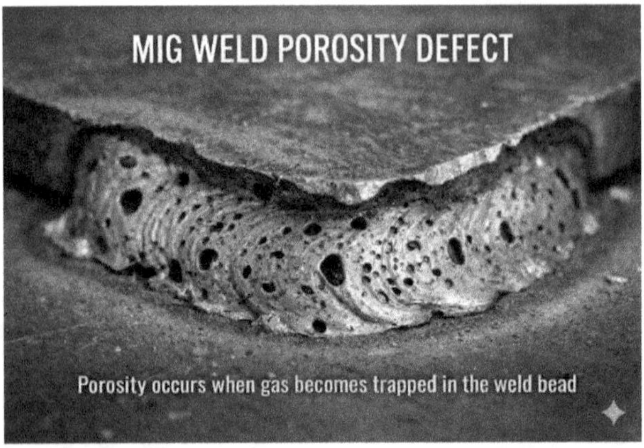

Cause: bad gas flow

Fix: check gas, block wind, clean metal

Spatter

Cause: wrong settings

Fix: adjust voltage/wire speed

Burn-Through

Cause: too much heat
Fix: lower voltage, move faster

Bird-Nesting

Cause: liner or roller issues
Fix: replace liner, adjust tension

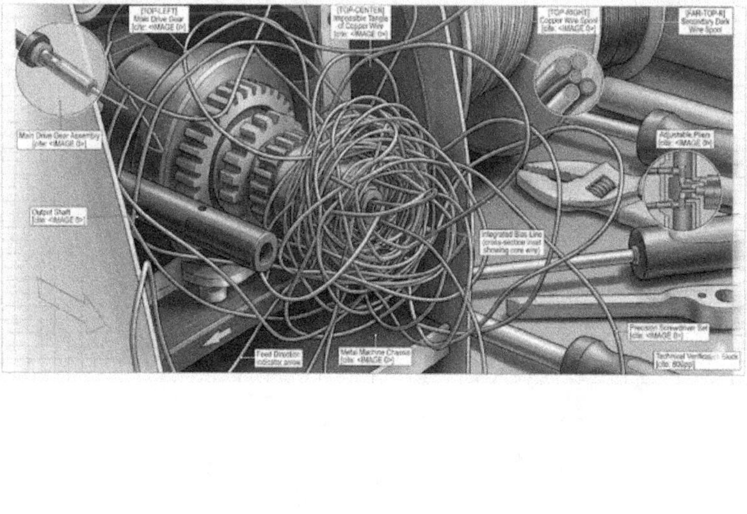

Why MIG Welding Matters

MIG is fast, clean, and versatile — the go-to process for fabrication and general repairs.

Summary

MIG and Flux-Core welding give you the speed, versatility, and ease of use needed for most shop and field work. Once you understand wire selection, machine setup, and proper technique, you can weld confidently in any position and troubleshoot problems quickly.

Chapter 5 — Knowledge Check

Chapter 5 Quiz

1. What is the main difference between MIG and Flux-Core welding?
A. MIG uses a foot pedal
B. MIG uses solid wire with external gas; Flux-Core uses tubular wire with flux
C. Flux-Core requires a spool gun
D. MIG can only be used outdoors

2. Which wire type is most commonly used for mild steel MIG welding?
A. ER4043
B. ER308L
C. ER70S-6
D. E71T-11

3. What stickout length is recommended for MIG welding?
A. ⅛"
B. ⅜"–½"
C. ¾"
D. 1"

4. What stickout length is recommended for Flux-Core welding?

A. ⅛"

B. ¼"

C. ⅜"

D. ¾"

5. Which shielding gas is most commonly used for MIG welding on mild steel?

A. 100% Argon

B. 75/25 Argon-CO_2

C. 100% CO_2 only

D. Helium

6. Which gun angle is correct for MIG welding?

A. 10–15° drag angle

B. 45° push angle

C. 10–15° push angle

D. Straight down

7. Which gun angle is correct for Flux-Core welding?

A. Push

B. Drag

C. Straight up

D. 90° perpendicular

8. What happens if you travel too fast while MIG welding?
A. Excessive buildup
B. Undercut and poor fusion
C. Overheating
D. Slag inclusions

9. Worm tracks in Flux-Core welding are caused by:
A. Too little heat
B. Too much gas flow
C. Wire running too hot or travel too slow
D. Using the wrong polarity

10. What is the purpose of watching the leading edge of the puddle?
A. To see the arc more clearly
B. To control bead width and penetration
C. To reduce spatter
D. To cool the weld faster

11. Which wire diameter is best for thin metal?
A. .023"
B. .030"
C. .035"
D. .045"

12. Why is Flux-Core preferred for outdoor welding?

A. It is cleaner than MIG

B. It requires no electricity

C. The flux creates its own shielding, unaffected by wind

D. It uses less wire

Answer Key

1. B — MIG uses solid wire with gas; Flux-Core uses tubular wire with flux.
2. C — ER70S-6 is the standard MIG wire for mild steel.
3. B — MIG stickout should be ⅜"–½".
4. D — Flux-Core uses a longer stickout of about ¾".
5. B — 75/25 Argon-CO_2 is the standard MIG gas for mild steel.
6. C — MIG uses a 10–15° push angle.
7. B — Flux-Core uses a drag angle.
8. B — Traveling too fast causes undercut and poor fusion.
9. C — Worm tracks come from running too hot or moving too slowly.
10. B — Watching the puddle controls bead shape and penetration.
11. A — .023" wire is best for thin metal.
12. C — Flux-Core's flux shielding is not blown away by wind.

CHAPTER 6 — STICK WELDING (SMAW)

What You Will Learn

Stick welding is the backbone of farm, field, and heavy-duty repair work. This chapter teaches you how Stick welding works, how to choose the right electrode, how to set polarity, and how to master the essential techniques needed for strong, reliable welds in any environment.

By the end of this chapter, you will understand:

- How Stick welding works and why it excels outdoors
- The differences between common electrode types (6011, 6013, 7018)
- How to select the correct polarity for each rod
- How to strike and maintain a stable arc
- How to control arc length, travel angle, and travel speed
- How to remove slag and inspect your weld
- How to troubleshoot common Stick welding problems

Stick welding is rugged, reliable, and extremely versatile — and this chapter gives you the skills to use it with confidence.

Understanding Stick Welding (SMAW)

This makes Stick welding ideal for:

- Outdoor work
- Dirty or rusty metal
- Heavy steel
- Long cable runs
- Farm and field repairs

Stick welding is less sensitive to wind, contamination, and imperfect conditions than MIG or TIG.

Electrode Types and Their Uses

Your document includes a table describing the most common rods. Here is the improved, reader-friendly version:

6011 — Deep Penetration, Tough Conditions

- Penetration: Deep
- Coating: Cellulosic
- Polarity: AC or DCEP
- Best For: Rusty, dirty, painted metal; outdoor repairs
- Notes: Excellent for farm and field work

6013 — Smooth, Easy-Running Rod

- Penetration: Medium
- Coating: Rutile
- Polarity: AC or DCEP
- Best For: Sheet metal, clean joints
- Notes: Easy slag removal, great for beginners

7018 — Strong, Low-Hydrogen Rod

- Penetration: Low to medium
- Coating: Low-hydrogen
- Polarity: DCEP
- Best For: Structural welds, clean steel
- Notes: Must be kept dry; produces strong, ductile welds

Choosing the right rod is one of the most important decisions in Stick welding.

Electrode Characteristics Table

Rod Type	Penetration	Coating Type	Polarity	Best For	Notes
6011	Deep	Cellulosic	AC or DCEP	Dirty/rusty metal, outdoor repair	Excellent for farm and field work
6013	Medium	Rutile	AC or DCEP	Sheet metal, clean joints	Smooth arc, easy slag removal
7018	Low to Medium	Low-Hydrogen	DCEP (DC+)	Structural welds, clean steel	Must be kept dry, strong welds

Polarity Settings

- 6010 — DCEP
- 6011 — AC or DCEP
- 6013 — AC or DCEP
- 7018 — DCEP

Using the wrong polarity can cause sticking, spatter, or poor penetration.

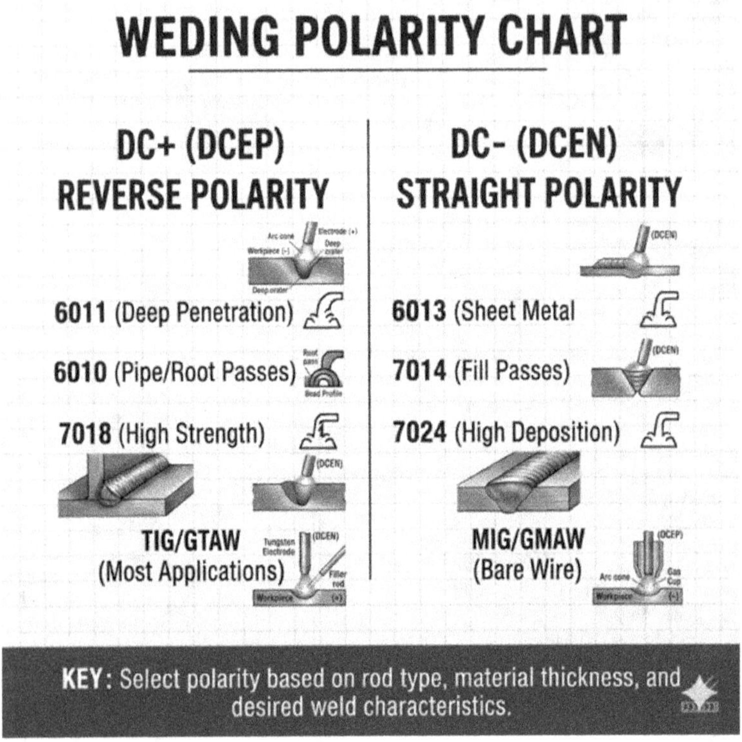

Striking the Arc

There are two ways to start a Stick weld:

Scratch Start

Drag the rod like striking a match.

Tap Start

Tap the rod to the metal and lift slightly.

Both methods require practice. Too much contact causes sticking; too little breaks the arc.

Stick Welding Technique

Arc Length

Keep the arc length equal to the rod diameter.
Too long = spatter and weak welds
Too short = sticking and unstable arc

Travel Angle

Use a 10–15° drag angle.
Stick welding is always a pull process.

Travel Speed

Watch the puddle — not the arc.
A good puddle wets into both sides of the joint.

Slag Removal

After each pass:

1. Let the weld cool slightly
2. Chip the slag
3. Brush clean
4. Inspect the bead

Slag must be removed completely before welding additional passes.

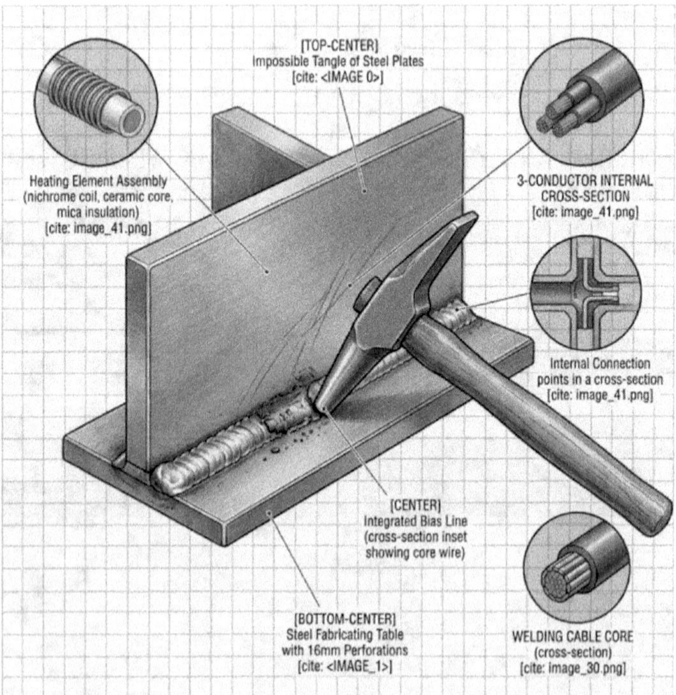

Common Stick Welding Problems

Sticking the Rod

Cause: Arc too short
Fix: Increase amperage or lengthen arc

Porosity

Cause: Moisture or wind
Fix: Dry rods, block wind, clean metal

Slag Inclusions

Cause: Wrong angle or speed
Fix: Maintain drag angle, slow down, clean between passes

Undercut

Cause: Too much heat
Fix: Reduce amperage, adjust travel speed

Stick welding is forgiving, but technique matters.

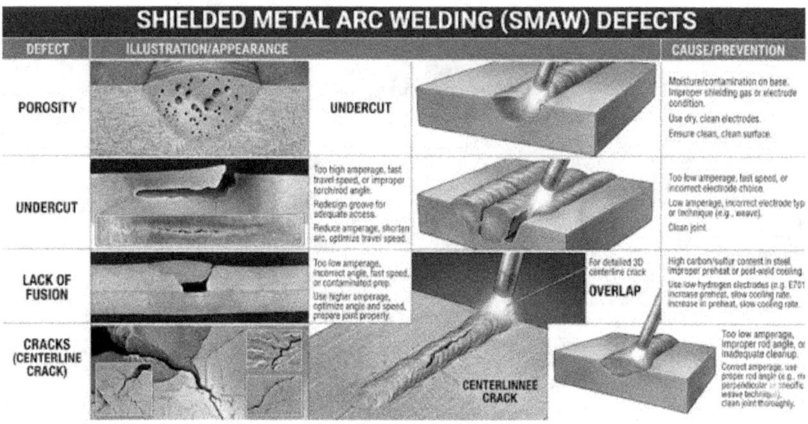

Why Stick Welding Matters

Stick welding is the workhorse of real-world repairs. Once you master rod selection, arc control, and slag management, you can weld confidently anywhere — from the shop to the field.

Chapter 6 — Knowledge Check

Chapter 6 Quiz

1. What creates the shielding gas and slag in Stick welding?
A. External gas cylinder
B. The flux coating on the electrode
C. A built-in gas reservoir
D. A cooling fan in the welder

2. Which electrode is best for welding through rust, paint, or dirty metal?
A. 6013
B. 7018
C. 6011
D. 7024

3. Which rod must be kept dry to prevent porosity and hydrogen cracking?
A. 6011
B. 6013
C. 7018
D. 6010

4. What is the correct travel angle for Stick welding?

A. 10–15° push angle

B. 10–15° drag angle

C. 45° straight-in angle

D. 90° perpendicular

5. What should your arc length be during Stick welding?

A. Equal to the rod diameter

B. 1 inch

C. As long as possible

D. Touching the metal at all times

6. What is the purpose of slag in Stick welding?

A. To cool the weld

B. To protect the molten metal from contamination

C. To increase amperage

D. To clean the metal automatically

7. What is the correct polarity for 7018 electrodes?

A. AC only

B. DCEN

C. DCEP

D. Either AC or DC

8. What causes a rod to stick to the metal?

A. Arc too long

B. Arc too short or amperage too low

C. Too much gas flow

D. Using the wrong wire diameter

9. Slag inclusions are usually caused by:

A. Too much amperage

B. Incorrect drag angle or failing to clean between passes

C. Using a rod that is too large

D. Welding too fast

10. Why is Stick welding ideal for outdoor and field repairs?

A. It requires no electricity

B. The flux shielding is not affected by wind

C. It produces no sparks

D. It uses the least amount of heat

11. What is the purpose of chipping and brushing after each pass?

A. To cool the weld

B. To remove slag and inspect the bead

C. To increase penetration

D. To prepare for painting

12. Which electrode is best for clean, smooth welds on clean steel?

A. 6011

B. 6013

C. 7018

D. 7024

Answer Key

1. B — The flux coating creates shielding gas and slag.
2. C — 6011 is ideal for dirty, rusty, or painted metal.
3. C — 7018 must be kept dry to prevent hydrogen issues.
4. B — Stick welding uses a 10–15° drag angle.
5. A — Arc length should equal the rod diameter.
6. B — Slag protects the molten weld from contamination.
7. C — 7018 runs on DCEP.
8. B — A rod sticks when the arc is too short or amperage too low.
9. B — Slag inclusions come from poor angle or not cleaning between passes.
10. B — Flux shielding is not blown away by wind.
11. B — Chipping and brushing remove slag and allow inspection.
12. C — 7018 produces strong, smooth welds on clean steel.

CHAPTER 7 — TIG WELDING (GTAW)

What You Will Learn

TIG welding is the most precise and controlled welding process. It produces clean, high-quality welds on steel, stainless steel, and aluminum — especially thin material. This chapter teaches you how TIG works, how to select the right tungsten, how to set up your machine, and how to master the hand-eye coordination needed for smooth, consistent welds.

By the end of this chapter, you will understand:

- How TIG welding works and why it produces the cleanest welds
- The differences between tungsten types and how to choose the right one
- How to set polarity, amperage, and gas flow
- How to prepare steel, stainless steel, and aluminum for TIG
- How to control torch angle, filler rod placement, and travel speed
- How to weld thin metal without warping or burn-through
- How to troubleshoot common TIG welding problems

TIG welding demands patience and precision — but once mastered, it becomes one of the most rewarding welding processes.

Understanding TIG Welding (GTAW)

This gives TIG unmatched control over:

- Heat input
- Puddle size
- Weld appearance
- Penetration
- Distortion

TIG is the preferred process for:

- Aluminum
- Stainless steel
- Thin metal
- Decorative welds
- Precision fabrication

Tungsten Types and Selection

Your document lists the most common tungsten electrodes. Here is the improved, reader-friendly version:

2% Thoriated (Red)

- Best for: Steel and stainless
- Polarity: DCEN
- Notes: Easy arc start, stable arc

2% Ceriated (Grey)

- Best for: Low-amp work, thin metal
- Polarity: DCEN
- Notes: Excellent for small parts and precision

2% Lanthanated (Blue or Gold)

- Best for: All-purpose TIG
- Polarity: AC or DC
- Notes: Great for aluminum and steel

Pure Tungsten (Green)

- Best for: Aluminum (older machines)
- Polarity: AC
- Notes: Balls easily, stable arc on AC

Choosing the right tungsten affects arc stability, penetration, and weld quality.

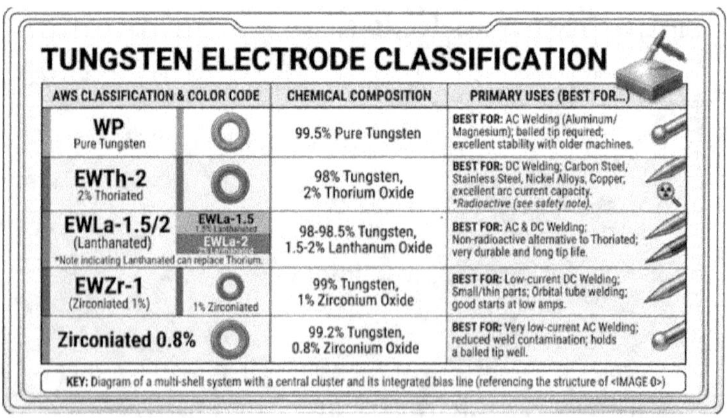

TUNGSTEN ELECTRODE CLASSIFICATION

AWS CLASSIFICATION & COLOR CODE		CHEMICAL COMPOSITION	PRIMARY USES (BEST FOR...)
WP Pure Tungsten		99.5% Pure Tungsten	**BEST FOR:** AC Welding (Aluminum/ Magnesium); balled tip required; excellent stability with older machines.
EWTh-2 2% Thoriated		98% Tungsten, 2% Thorium Oxide	**BEST FOR:** DC Welding; Carbon Steel, Stainless Steel, Nickel Alloys, Copper; excellent arc current capacity. **Radioactive (see safety note).*
EWLa-1.5/2 (Lanthanated) *Note indicating Lanthanated can replace Thorium.	EWLa-1.5 1.5% Lanthanated EWLa-2	98-98.5% Tungsten, 1.5-2% Lanthanum Oxide	**BEST FOR:** AC & DC Welding; Non-radioactive alternative to Thoriated; very durable and long tip life.
EWZr-1 (Zirconiated 1%)	1% Zirconiated	99% Tungsten, 1% Zirconium Oxide	**BEST FOR:** Low-current DC Welding; Small/thin parts; Orbital tube welding; good starts at low amps.
Zirconiated 0.8%		99.2% Tungsten, 0.8% Zirconium Oxide	**BEST FOR:** Very low-current AC Welding; reduced weld contamination; holds a balled tip well.

KEY: Diagram of a multi-shell system with a central cluster and its integrated bias line (referencing the structure of <IMAGE 0>)

Machine Settings

Polarity

- Steel / Stainless: DCEN
- Aluminum: AC

Amperage

Match amperage to metal thickness.
General rule: 1 amp per .001" of thickness

Gas Flow

- 15–20 CFH for most TIG work
- 100% Argon is standard

Cup Size

- #6–#8 for most steel
- #8–#12 for aluminum

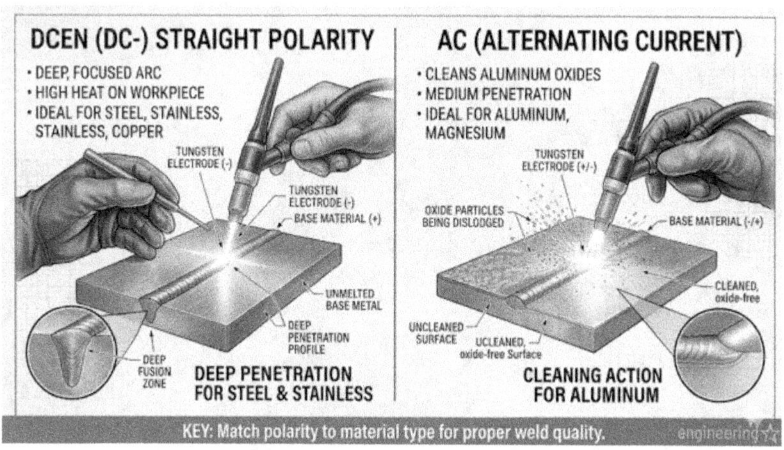

Preparing Metal for TIG Welding

TIG is extremely sensitive to contamination.

Steel

- Remove rust, paint, oil, and mill scale
- Wipe with acetone

Stainless Steel

- Clean with acetone
- Use stainless-only brushes
- Avoid cross-contamination

Aluminum

- Remove oxide layer with stainless brush
- Wipe with acetone
- Keep filler rod clean

Cleanliness is the foundation of good TIG welding.

TIG Welding Technique

Torch Angle

Use a 10–15° push angle.
Never drag the torch.

Arc Length

Keep the tungsten ⅛" or less from the puddle.
Too long = unstable arc
Too short = contamination

Filler Rod Technique

Dip the filler into the leading edge of the puddle.
Do NOT melt the rod with the arc.

Travel Speed

Move steadily and watch the puddle wet into both sides of the joint.

Foot Pedal Control

Feather the pedal to control heat.

- More heat to start
- Less heat as the part warms up

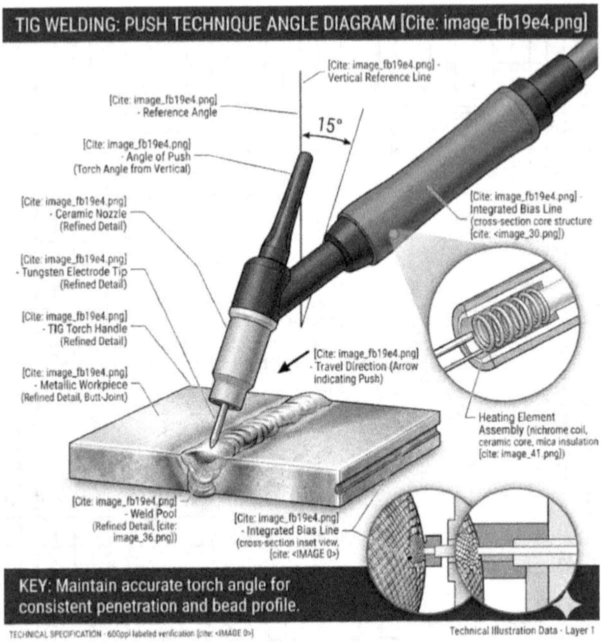

Welding Aluminum with TIG

Aluminum is challenging but rewarding.

- Use AC
- Let cleaning action remove oxide
- Keep tungsten rounded
- Move quickly

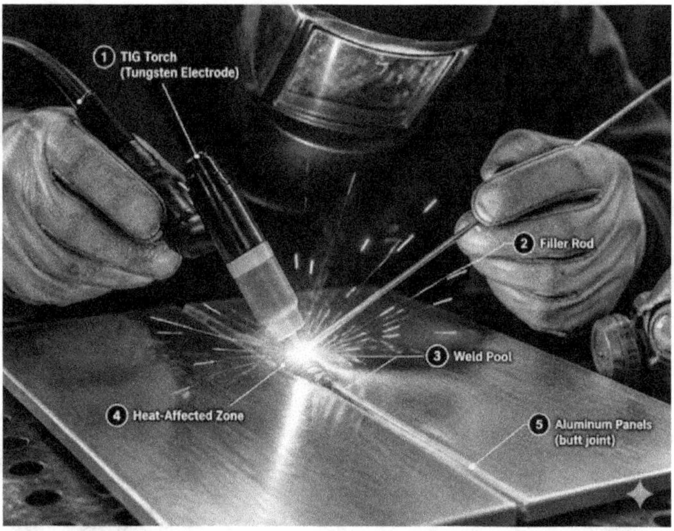

Welding Steel with TIG

Steel is the easiest TIG material.

- Use DCEN
- Keep tungsten sharp
- Maintain short arc
- Add filler consistently

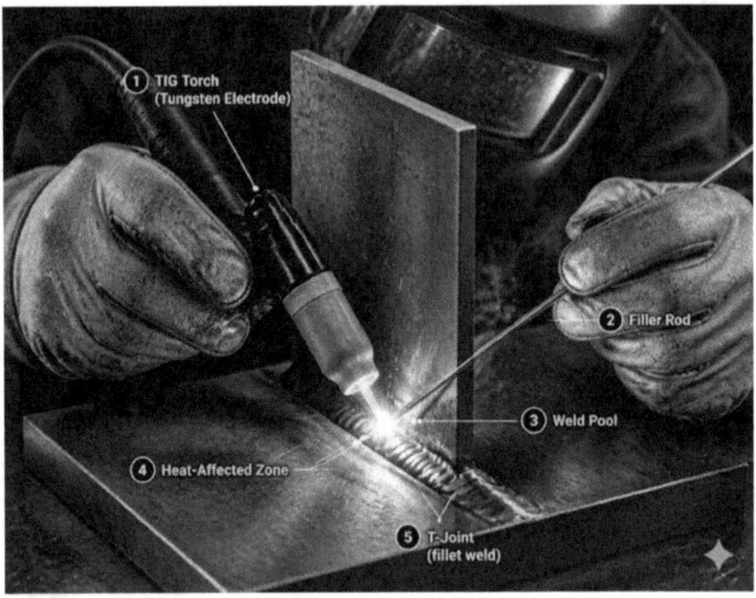

Welding Stainless Steel with TIG

Stainless holds heat longer and warps easily.

- Use DCEN
- Keep heat low
- Move steadily
- Use chill blocks if needed

Common TIG Problems and Fixes

Contaminated Tungsten

Cause: touching puddle or filler
Fix: regrind tungsten

Porosity

Cause: bad gas flow or dirty metal
Fix: clean metal, increase gas flow

Overheating

Cause: too much heat
Fix: ease off pedal, move faster

Lack of Fusion

Cause: Low heat or moving too fast
Fix: Increase amperage, slow down

Wandering Arc

Cause: dull tungsten
Fix: sharpen tungsten

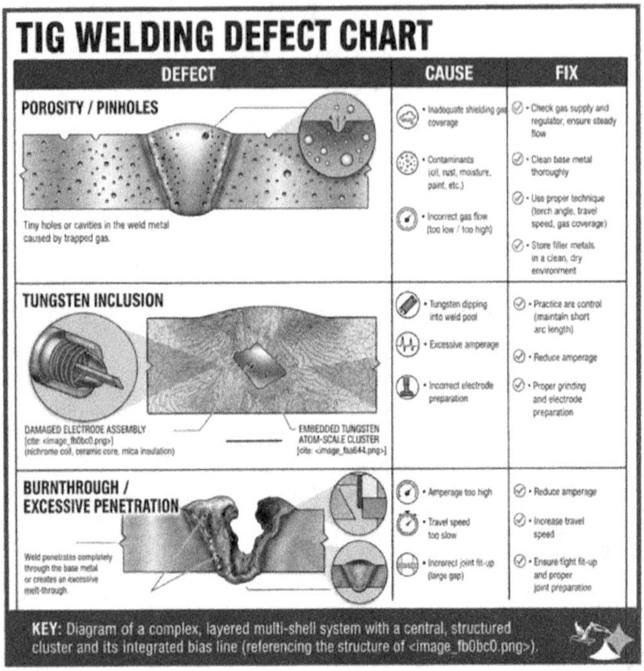

Why TIG Welding Matters

It is the process of choice for:

- High-end fabrication
- Automotive and motorcycle work
- Stainless and aluminum projects
- Thin metal
- Decorative welds

TIG welding rewards patience, control, and attention to detail — and this chapter gives you the foundation to build those skills.

Chapter 7 — Knowledge Check

Chapter 7 Quiz

1. What makes TIG welding different from MIG and Stick welding?
A. It uses a consumable wire electrode
B. It uses a non-consumable tungsten electrode and separate filler rod
C. It requires no shielding gas
D. It is only used outdoors

2. Which polarity is used for TIG welding aluminum?
A. DCEN
B. DCEP
C. AC
D. Either AC or DC

3. Which tungsten type is considered the best all-purpose option for both AC and DC welding?
A. Pure (Green)
B. 2% Thoriated (Red)
C. 2% Lanthanated (Blue/Gold)
D. 2% Ceriated (Grey)

4. What is the recommended gas for TIG welding steel, stainless, and aluminum?

A. CO_2

B. 75/25 Argon-CO_2

C. 100% Argon

D. Helium only

5. What is the correct torch angle for TIG welding?

A. 10–15° push angle

B. 10–15° drag angle

C. 45° straight-in

D. 90° perpendicular

6. What happens if your tungsten touches the puddle or filler rod?

A. The weld becomes stronger

B. The tungsten becomes contaminated

C. The arc becomes hotter

D. Nothing — it is normal

7. Why is aluminum more challenging to TIG weld?

A. It is magnetic

B. It has low thermal conductivity

C. Its oxide layer melts at a higher temperature than the base metal

D. It requires no cleaning

8. What is the general amperage rule for TIG welding?
A. 10 amps per .001"
B. 1 amp per .001"
C. 5 amps per .001"
D. 20 amps per .001"

9. What is the correct way to add filler rod in TIG welding?
A. Melt the rod directly with the arc
B. Dip the rod into the leading edge of the puddle
C. Hold the rod still and move the torch only
D. Push the rod under the tungsten

10. What causes porosity in TIG welds?
A. Too much amperage
B. Dirty metal or poor gas coverage
C. Using a cup that is too large
D. Moving too slowly

11. What is the purpose of AC balance when welding aluminum?
A. To increase amperage
B. To control cleaning action vs. penetration
C. To reduce gas flow
D. To sharpen the tungsten

12. What is the most common cause of overheating and warping when TIG welding thin metal?
A. Too little heat
B. Too much heat input
C. Using a small cup
D. Using a foot pedal

Answer Key

1. B — TIG uses a non-consumable tungsten electrode and separate filler rod.
2. C — Aluminum is welded on AC.
3. C — 2% Lanthanated is an excellent all-purpose tungsten.
4. C — TIG welding uses 100% Argon.
5. A — TIG uses a 10–15° push angle.
6. B — Touching the puddle contaminates the tungsten.
7. C — Aluminum's oxide layer melts at a higher temperature than the base metal.
8. B — The general rule is 1 amp per .001" of thickness.
9. B — Add filler to the leading edge of the puddle.
10. B — Porosity comes from contamination or poor gas coverage.
11. B — AC balance controls cleaning action vs. penetration.
12. B — Too much heat causes warping and overheating.

CHAPTER 8 — METAL PREPARATION & JOINT DESIGN

What You Will Learn

Strong welds start long before you strike an arc. This chapter teaches you how to prepare metal correctly, how to choose the right joint type, and how to ensure proper fit-up for clean, reliable welds. Good preparation prevents porosity, cracking, lack of fusion, and distortion — the most common causes of weld failure.

By the end of this chapter, you will understand:

- How to clean and prepare steel, stainless steel, and aluminum
- How to remove rust, paint, mill scale, and contaminants
- How to bevel edges for full penetration
- How to select the correct joint type for your project
- How to achieve proper fit-up and gap spacing
- How to use tacks, clamps, and fixtures to control alignment
- How joint design affects strength, penetration, and distortion

This chapter gives you the foundation for consistent, professional-quality welds.

Why Metal Preparation Matters

Even the best welder cannot overcome:

- Dirty metal
- Poor fit-up
- Incorrect joint design
- Improper beveling
- Gaps that are too wide or too tight

Preparation is not optional — it is part of the welding process.

Cleaning and Preparing Metal

Different metals require different preparation methods.

Preparing Mild Steel

Mild steel is forgiving, but still requires clean surfaces.

Remove:

- Rust
- Paint
- Oil and grease
- Mill scale
- Dirt and moisture

Tools to Use

- Flap disc
- Grinding wheel
- Wire wheel
- Acetone or degreaser

Clean metal produces cleaner puddles, better penetration, and stronger welds.

Preparing Stainless Steel

Stainless steel shows contamination easily and can trap impurities.

Preparation Steps

- Clean with acetone
- Use stainless-only brushes and wheels
- Avoid touching with bare hands (oils cause porosity)
- Remove heat tint between passes if required

Cross-contamination from carbon steel tools can ruin a stainless weld.

Preparing Aluminum

Preparation Steps

- Remove oxide layer with a stainless brush
- Wipe with acetone
- Keep filler rod clean
- Ensure tight fit-up (aluminum expands quickly)

Clean aluminum welds beautifully — dirty aluminum does not weld at all.

Beveling and Edge Preparation

Beveling allows full penetration on thicker metal.

When to Bevel

- Steel thicker than 1/8"
- Structural welds
- Multi-pass welds
- Joints requiring maximum strength

Common Bevel Types

- Single bevel
- Double bevel
- V-groove
- U-groove

Bevels create space for the weld to penetrate deeply and fuse properly.

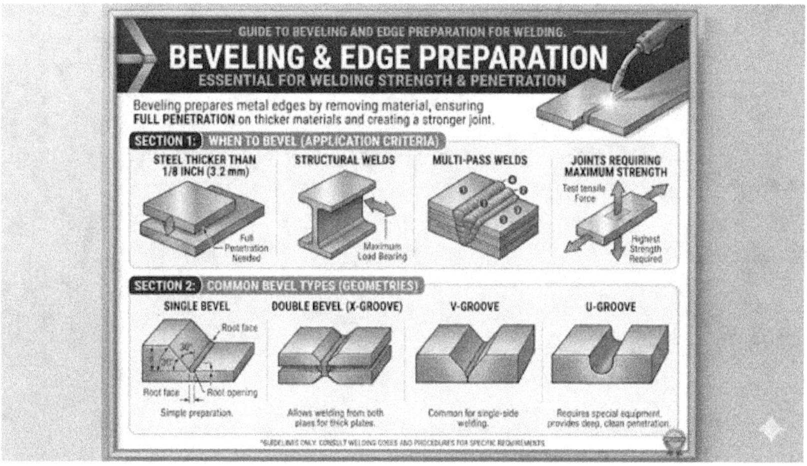

Joint Types and Their Uses

Butt Joint

Two pieces end-to-end

- Best for: Plate, tubing, structural work
- Notes: Often requires beveling

Lap Joint

One piece overlaps another.

T-Joint

One piece meets another at 90°

- Best for: Frames, brackets, fabrication
- Notes: Watch for undercut at the toes

Corner Joint

Two pieces meet at the edge

- Best for: Boxes, frames, enclosures
- Notes: Heat control is critical

Edge Joint

Two edges side-by-side

- Best for: Thin metal
- Notes: Low heat input required

Choosing the right joint design affects strength, appearance, and ease of welding.

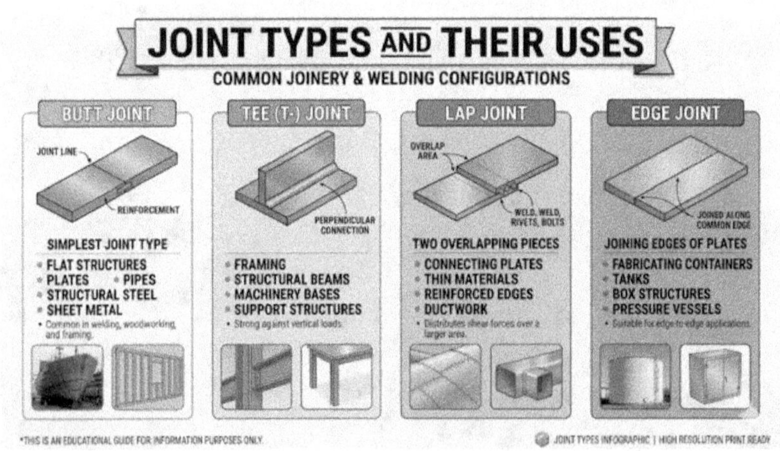

Fit-Up and Alignment

Good fit-up makes welding easier and stronger.

Key Principles

- Gaps should be consistent
- Edges should be flush
- Tacks should be evenly spaced
- Clamps should hold pieces firmly
- Heat should be balanced to prevent distortion

Tack Welding

Use tacks to:

- Hold alignment
- Prevent warping
- Maintain gap spacing
- Lock the joint before welding

More tacks = less distortion.

Controlling Distortion

Distortion Control Techniques

- Use more tacks
- Weld in short sections
- Alternate sides
- Clamp tightly
- Allow cooling time
- Use chill bars or backing plates

Distortion is predictable — and preventable.

Why Joint Design Matters

Joint design affects:

- Penetration
- Strength
- Heat input
- Distortion
- Weld appearance
- Number of passes required

A well-designed joint makes welding easier and produces stronger results.

Summary

Metal preparation and joint design are the foundation of strong, clean welds. When you start with clean metal, proper bevels, and tight fit-up, every welding process becomes easier — and your results become more consistent and professional.

Chapter 8 — Knowledge Check

Chapter 8 Quiz

1. Why is metal preparation essential before welding?
A. It makes the weld look shiny
B. It prevents porosity, cracking, and lack of fusion
C. It increases amperage
D. It eliminates the need for clamps

2. What must be removed from mild steel before welding?
A. Only paint
B. Only rust
C. Rust, paint, oil, grease, and mill scale
D. Nothing — mild steel can be welded as-is

3. Why must stainless steel be cleaned with stainless-only brushes?
A. Stainless brushes are cheaper
B. Carbon steel brushes contaminate the weld
C. Stainless brushes create deeper penetration
D. Stainless brushes remove slag

4. What makes aluminum preparation unique?
A. It requires no cleaning
B. Its oxide layer melts at a higher temperature than the base metal
C. It is magnetic
D. It does not expand when heated

5. When should metal be beveled?

A. Only on thin sheet metal

B. Only when welding aluminum

C. When welding steel thicker than 1/8" or when full penetration is needed

D. Never — beveling weakens the joint

6. Which joint type is best for sheet metal and automotive work?

A. Butt joint

B. Lap joint

C. T-joint

D. Corner joint

7. Which joint type is commonly used for frames and brackets?

A. Edge joint

B. Lap joint

C. T-joint

D. Corner joint

8. What is the purpose of tack welds?

A. To finish the weld faster

B. To hold alignment, prevent warping, and maintain gap spacing

C. To increase amperage

D. To remove slag

9. What causes distortion during welding?
A. Low amperage
B. Uneven heat input and expansion
C. Using too many clamps
D. Welding too slowly

10. Which technique helps control distortion?
A. Welding the entire joint in one long pass
B. Using more tacks and alternating sides
C. Increasing heat input
D. Leaving the joint unclamped

11. What is the main purpose of joint design?
A. To make the weld look decorative
B. To determine strength, penetration, and heat distribution
C. To reduce the need for cleaning
D. To eliminate the need for filler metal

12. Why is consistent gap spacing important?
A. It makes the weld quieter
B. It ensures predictable penetration and bead shape
C. It reduces gas usage
D. It prevents slag formation

Answer Key

1. B — Proper preparation prevents porosity, cracking, and lack of fusion.
2. C — Mild steel must be cleaned of rust, paint, oil, grease, and mill scale.
3. B — Carbon steel brushes contaminate stainless steel.
4. B — Aluminum's oxide layer melts at a higher temperature than the base metal.
5. C — Bevel when welding steel thicker than 1/8" or when full penetration is needed.
6. B — Lap joints are ideal for sheet metal and automotive work.
7. C — T-joints are common for frames and brackets.
8. B — Tacks maintain alignment, spacing, and prevent warping.
9. B — Distortion is caused by uneven heat input and expansion.
10. B — More tacks and alternating sides help control distortion.
11. B — Joint design determines strength, penetration, and heat distribution.
12. B — Consistent gaps ensure predictable penetration and bead shape.

CHAPTER 9 — WELDING POSITIONS & TECHNIQUES

What You Will Learn

Welders rarely work on perfectly flat tables. Real-world welding requires skill in multiple positions — horizontal, vertical, and overhead — each with its own challenges. This chapter teaches you how to control the puddle, manage heat, and maintain proper technique in every welding position.

By the end of this chapter, you will understand:

- The four primary welding positions
- How gravity affects the puddle in each position
- How to adjust travel speed, angle, and heat input
- How to weld vertical-up, vertical-down, and overhead
- How to choose the right weave pattern
- How to maintain control and avoid defects in difficult positions

Mastering welding positions makes you a more capable, versatile welder — whether you're repairing equipment, fabricating frames, or working in the field.

The Four Primary Welding Positions

1. Flat Position (1G / 1F)

- Easiest position
- Puddle is stable
- Ideal for beginners
- Allows higher heat and faster travel

2. Horizontal Position (2G / 2F)

- Gravity pulls the puddle downward
- Requires tighter control
- Slight upward torch/gun angle helps support the puddle

3. Vertical Position (3G / 3F)

- Welds run up or down the plate
- Heat control is critical
- Vertical-up gives deeper penetration
- Vertical-down is useful for thin metal

4. Overhead Position (4G / 4F)

- Most challenging
- Puddle wants to fall
- Requires lower heat and tight control
- Keep puddle small and fast

Each position demands adjustments in technique, heat, and movement.

Flat Position Technique

The flat position is the foundation of all welding.

Key Points

- Use normal heat settings
- Maintain steady travel speed
- Watch the puddle wet into both sides
- Use small, controlled motions

Flat welding builds confidence and consistency.

Horizontal Position Technique

Gravity pulls the molten metal downward, so adjustments are needed.

Key Points

- Angle the gun/torch slightly upward
- Reduce heat slightly
- Move steadily to avoid sagging
- Keep the puddle small and controlled

Horizontal welds require balance between heat and travel speed.

Vertical Welding Techniques

Vertical-Up (3G Up)

Best For

- Thick metal
- Structural welds
- Strong, deep penetration

Technique

- Use a tight weave or “Christmas tree” pattern
- Pause at the toes for fusion
- Move upward slowly
- Use lower heat than flat position

Vertical-up produces strong welds but requires patience and control.

Vertical-Down (3G Down)

Best For

- Thin metal
- Sheet metal
- Automotive work

Technique

- Move quickly
- Use a slight push angle
- Keep heat low
- Avoid weaving

Vertical-down is fast and clean but not ideal for thick steel.

Overhead Welding Technique

Overhead welding is challenging because the puddle wants to fall.

Key Points

- Reduce heat
- Keep puddle small
- Maintain tight arc length
- Move quickly and steadily
- Use short, controlled motions

Good overhead welds come from discipline and practice.

Weave Patterns

Straight Stringer

- Best for beginners
- Strongest weld
- Least heat input

Small Weave

- Useful for filling wider joints
- Helps control puddle in vertical-up

Crescent / "Christmas Tree"

- Common in vertical-up
- Pause at the toes for fusion

Z-Weave

- Used for heavy plate
- Helps distribute heat

Weaves should be controlled and purposeful — not wide or sloppy.

Heat Control in All Positions

Heat affects:

- Penetration
- Puddle size
- Distortion
- Weld appearance

General Rules

- Lower heat for vertical and overhead
- Higher heat for flat and horizontal
- Move faster when metal heats up
- Watch the puddle, not the arc

Heat management is the key to mastering all positions.

Common Problems in Out-of-Position Welding

Sagging / Drooping Bead

Cause: Too much heat
Fix: Lower amperage, move faster

Undercut

Cause: Wrong angle or too fast
Fix: Adjust angle, slow down

Cold Lap

Cause: Too little heat
Fix: Increase amperage, improve fusion

Excessive Spatter

Cause: Long arc length
Fix: Shorten arc, maintain control

Why Position Mastery Matters

Mastering all positions allows you to:

- Weld equipment in place
- Repair farm machinery
- Fabricate frames and structures
- Work confidently in the field
- Produce strong welds in any orientation

This chapter builds the skills needed for real-world welding success.

Chapter 9 — Knowledge Check

Chapter 9 Quiz

1. What is the easiest welding position for beginners?

A. Horizontal

B. Vertical-up

C. Flat

D. Overhead

2. Why is horizontal welding more challenging than flat welding?

A. It requires more amperage

B. Gravity pulls the puddle downward

C. It uses a different filler metal

D. It requires no torch angle

3. Which welding position requires the puddle to be kept very small to prevent it from falling?

A. Flat

B. Horizontal

C. Vertical-down

D. Overhead

4. Vertical-up welding is best for:
A. Thin sheet metal
B. Decorative welds only
C. Thick metal and structural welds
D. Aluminum only

5. Vertical-down welding is ideal for:
A. Thick plate
B. Structural welds
C. Thin metal and sheet metal
D. Overhead joints

6. What is the purpose of pausing at the toes during a vertical-up weave?
A. To cool the weld
B. To improve fusion on both sides
C. To increase travel speed
D. To reduce gas usage

7. What is the biggest challenge in overhead welding?
A. Lack of penetration
B. Puddle falling due to gravity
C. Excessive gas flow
D. Tungsten contamination

8. Which weave pattern is commonly used for vertical-up welding?

A. Straight stringer

B. Circular weave

C. "Christmas tree" or triangle pattern

D. Spiral weave

9. What causes undercut in out-of-position welding?

A. Too little heat

B. Too much heat or incorrect angle

C. Using a small cup

D. Welding too slowly

10. How should heat be adjusted for vertical and overhead welding?

A. Increase heat significantly

B. Use the same heat as flat welding

C. Lower heat to maintain control

D. Turn off gas flow

11. What is the main advantage of mastering all welding positions?

A. Faster welds

B. Ability to weld equipment and structures in place

C. Reduced need for clamps

D. Eliminates the need for joint preparation

12. What should you watch while welding in any position?
A. The arc only
B. The puddle and how it wets into the joint
C. The sparks
D. The wire spool

Answer Key

1. C — Flat is the easiest and most stable position.
2. B — Gravity pulls the puddle downward in horizontal welding.
3. D — Overhead welding requires a very small, controlled puddle.
4. C — Vertical-up is ideal for thick metal and structural welds.
5. C — Vertical-down is best for thin metal and sheet metal.
6. B — Pausing at the toes improves fusion on both sides.
7. B — Gravity causes the puddle to fall in overhead welding.
8. C — The "Christmas tree" pattern is common for vertical-up.
9. B — Undercut comes from too much heat or incorrect angle.
10. C — Lower heat is needed for vertical and overhead control.
11. B — Position mastery allows welding equipment and structures in place.
12. B — Watching the puddle ensures proper fusion and bead shape.

CHAPTER 10 — FABRICATION SKILLS

What You Will Learn

Welding is only one part of metalwork. Fabrication is the complete process of turning raw metal into a finished project — measuring, cutting, fitting, tacking, welding, and finishing. This chapter teaches you the practical shop skills that make your welds straighter, your projects stronger, and your work more professional.

By the end of this chapter, you will understand:

- How to measure and lay out metal accurately
- How to choose the right cutting tool for the job
- How to fit, square, and align metal pieces
- How to use tack welds to control movement
- How to manage heat and prevent distortion
- How to build simple projects from start to finish
- How to grind, clean, and finish your work

Fabrication is where welding becomes building — and mastering these skills unlocks the ability to create almost anything.

Measuring and Layout

Accuracy is the foundation of every good project. A weld can be perfect, but if the pieces are cut wrong or out of square, the project will fail.

Essential Layout Tools

- Tape measure
- Speed square
- Combination square
- Scribe
- Soapstone or paint marker
- Center punch

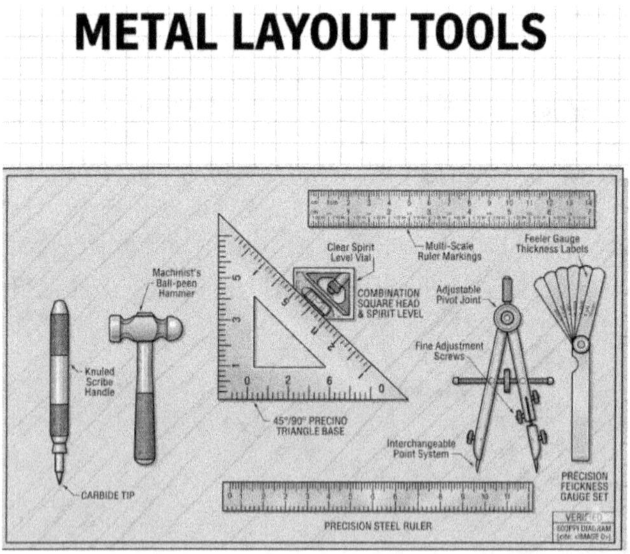

Layout Principles

- Mark all cut lines clearly
- Use a center punch for holes
- Double-check measurements before cutting
- Keep tools clean and accurate

Good layout saves time, prevents mistakes, and ensures clean fit-up.

Cutting Metal

Different cutting tools excel at different tasks. Choosing the right one improves accuracy and reduces cleanup.

Angle Grinder

- Portable and versatile
- Good for trimming, small cuts, and cleanup

Chop Saw

- Fast and accurate
- Ideal for repetitive cuts and square ends

Plasma Cutter

- Clean, fast cuts on thick or thin metal
- Excellent for curves and irregular shapes

Oxy-Acetylene Torch

- Great for thick steel
- Useful for heating, bending, and rough cutting

Each tool has strengths — use the one that matches the job.

Angle Grinder Chop Saw Plasma Cutter Oxy-Acetylene To

Fitting and Squaring

Good fit-up makes welding easier, cleaner, and stronger.

Clamps and Magnets

Use clamps to hold pieces securely:

- C-clamps
- Locking pliers
- Bar clamps
- Magnetic squares

Checking for Square

Use:

- Speed square
- Carpenter's square
- Diagonal measurements (measure corner-to-corner on frames)

If the diagonals match, the frame is square.

Tack Welding for Alignment

Tack welds lock the project together before final welding.

Good Tack Welds Should:

- Be small but strong
- Hold alignment
- Prevent distortion
- Be placed evenly

Where to Place Tacks

- Opposite corners
- Midpoints on long joints
- Anywhere movement is likely

More tacks = less distortion.

Controlling Heat and Distortion

Heat causes metal to expand and contract. If not controlled, your project will twist or pull out of square.

Ways to Reduce Distortion

- Use more tacks
- Weld in short sections
- Alternate sides of the joint
- Clamp tightly
- Use chill bars or backing plates

Heat management is one of the most important fabrication skills.

Building Simple Projects

Every project — from a small bracket to a full trailer — follows the same pattern:

1. Measure
2. Cut
3. Fit
4. Tack
5. Check alignment
6. Weld
7. Clean and finish

Master this sequence and you can build almost anything.

Finishing Work

After welding, clean and finish the project.

Grinding

- Flap discs for smoothing
- Grinding wheels for heavy removal
- Wire wheels for cleaning

Coating

Protect metal with:

- Paint
- Primer
- Powder coating
- Oil (temporary protection)

Finishing improves appearance and prevents rust.

Why Fabrication Skills Matter

Fabrication is where welding becomes useful. With strong fabrication skills, you can build:

- Gates
- Tables
- Brackets
- Frames
- Trailers
- Equipment repairs

- Custom projects

When you can measure accurately, cut cleanly, fit tightly, and weld confidently, you can turn raw metal into anything you imagine.

Chapter 10 — Knowledge Check

Chapter 10 Quiz

1. Why is accurate layout important in fabrication?
A. It makes the weld shinier
B. It prevents mistakes and ensures proper fit-up
C. It reduces gas usage
D. It eliminates the need for clamps

2. Which tool is best for fast, repetitive straight cuts?
A. Angle grinder
B. Chop saw
C. Oxy-acetylene torch
D. File

3. What is the purpose of a center punch?
A. To remove slag
B. To mark hole locations
C. To clean welds
D. To measure angles

4. Which cutting tool is best for curves and irregular shapes?
A. Chop saw
B. Plasma cutter
C. Speed square
D. Wire wheel

5. What does matching diagonal measurements on a frame indicate?

A. The metal is too thin

B. The frame is square

C. The welds are too cold

D. The clamps are loose

6. Why are tack welds important?

A. They replace final welds

B. They hold alignment and prevent distortion

C. They cool the metal

D. They eliminate the need for clamps

7. Which method helps reduce distortion?

A. Welding the entire joint in one long pass

B. Using more tacks and alternating sides

C. Increasing heat

D. Leaving the joint unclamped

8. What is the correct order for building a simple project?

A. Weld → Cut → Tack → Paint

B. Fit → Paint → Weld → Grind

C. Measure → Cut → Fit → Tack → Weld → Finish

D. Grind → Weld → Measure → Paint

9. Which tool is best for smoothing welds?

A. Wire wheel

B. Flap disc

C. Chop saw

D. Center punch

10. Why do fabrication skills matter?
A. They reduce electricity usage
B. They allow you to build strong, accurate projects
C. They eliminate the need for welding
D. They are only useful for professionals

Answer Key

1. B — Accurate layout prevents mistakes and ensures proper fit-up.
2. B — Chop saws are ideal for fast, repetitive straight cuts.
3. B — A center punch marks hole locations.
4. B — Plasma cutters excel at curves and irregular shapes.
5. B — Matching diagonals means the frame is square.
6. B — Tacks hold alignment and prevent distortion.
7. B — More tacks and alternating sides reduce distortion.
8. C — Measure → Cut → Fit → Tack → Weld → Finish is the correct sequence.
9. B — Flap discs are best for smoothing welds.
10. B — Fabrication skills allow you to build strong, accurate projects.

CHAPTER 11 — FARM & FIELD WELDING

What You Will Learn

Farm and field welding is real-world welding — dirty metal, bad weather, awkward positions, and equipment that needs to be fixed *right now*. This chapter teaches you how to weld confidently outside the shop, using portable power sources, practical field prep techniques, and reinforcement methods that keep equipment running.

By the end of this chapter, you will understand:

- How outdoor conditions affect welding
- How to choose and manage field power sources
- How to prepare metal in less-than-ideal environments
- How to perform common farm and equipment repairs
- How to reinforce worn or cracked components
- How to weld safely and effectively in the field

Farm and field welding is about problem-solving — and this chapter gives you the skills to handle whatever breaks.

Outdoor Challenges

Welding outdoors introduces variables you don't face in a shop.

Wind

- Blows away shielding gas
- Causes porosity
- Makes MIG and TIG difficult

Solutions:

- Use Stick or Flux-Core
- Block wind with shields or equipment
- Increase gas flow if using MIG

Weather

- Cold metal cracks more easily
- Rain and moisture cause porosity
- Heat affects distortion

Solutions:

- Preheat thick or cold metal
- Keep rods and wire dry
- Avoid welding in rain for safety

Dirty or Rusty Metal

Farm equipment is rarely clean.

Solutions:

- Use 6011 or Flux-Core
- Grind aggressively
- Remove grease and oil with solvent

Field welding is about adapting to the environment.

Power Sources

Generator Welders

- Provide both power and welding output
- Ideal for remote locations
- Can run grinders and lights

Engine-Driven Welders

- Gas or diesel powered
- High amperage output
- Excellent for Stick and Flux-Core

Inverter Welders with Generators

- Lightweight
- Efficient
- Good for MIG or TIG if power is stable

Battery-Powered Welders (Emerging Tech)

- Limited duty cycle
- Useful for quick repairs

Choose the power source that matches the job and location.

Metal Prep in the Field

Field prep is rarely perfect — but it must be *good enough* to weld safely.

Cleaning

- Remove rust, paint, and dirt
- Use a grinder or wire wheel
- Wipe with solvent if needed

Beveling

- Use a grinder to open cracks
- Bevel thick metal for penetration

Fit-Up

- Use clamps, chains, or ratchet straps
- Align parts as close as possible
- Tack heavily to prevent movement

Preheating

Useful for:

- Thick steel
- Cold weather
- High-carbon components

A propane torch works well in the field.

Common Repairs

Farm and equipment repairs follow predictable patterns.

Cracked Brackets

- Grind out the crack
- Bevel edges
- Weld with 6011 root + 7018 cap
- Add gussets if needed

Broken Hitches

- Clean thoroughly
- Use multiple passes
- Reinforce with plate

Loader Arms & Buckets

- Remove dirt and grease
- Use heavy rod (7018 or 7024)
- Add wear plates

Implements & Frames

- Straighten bent metal with heat
- Weld slowly to avoid pulling out of square
- Add braces or angle iron

Field repairs must be strong, not pretty.

Reinforcement Techniques

Reinforcement prevents repeat failures.

Gussets

- Add strength to corners and brackets
- Use triangular plates

Fish Plates

- Overlap cracks with a plate
- Round corners to prevent new cracks

Sleeves

- Slide tubing over damaged sections
- Weld around both ends

Backing Plates

- Add thickness behind thin or worn metal

Reinforcement is cheap insurance on hard-working equipment.

Why Farm & Field Welding Matters

Farm and field welding keeps equipment running when downtime is not an option.
With strong field welding skills, you can:

- Repair equipment on-site
- Fix cracks before they spread
- Reinforce weak points
- Save time and money
- Keep operations moving

This chapter gives you the confidence to weld anywhere — not just in the shop.

Chapter 11 — Knowledge Check

Chapter 11 Quiz

1. Why is MIG welding difficult outdoors?
A. It overheats easily
B. Wind blows away shielding gas
C. It requires too much power
D. It cannot weld steel

2. Which welding process is most reliable in windy conditions?
A. TIG
B. MIG
C. Stick
D. Aluminum spool gun

3. What is the main advantage of engine-driven welders?
A. They are silent
B. They provide high amperage anywhere
C. They require no fuel
D. They only work indoors

4. What should you do before welding a crack in a bracket?
A. Weld directly over the crack
B. Grind out and bevel the crack
C. Cool the metal with water
D. Add paint first

5. What is a fish plate used for?

A. Decorative welding

B. Reinforcing cracks

C. Cutting metal

D. Removing slag

6. Why is preheating useful in field welding?

A. It makes the weld shiny

B. It reduces cracking in thick or cold metal

C. It increases spatter

D. It eliminates the need for beveling

7. What tool is most useful for field metal prep?

A. File

B. Angle grinder

C. Chop saw

D. Bench sander

8. What is the purpose of gussets?

A. To reduce amperage

B. To strengthen corners and brackets

C. To cool the weld

D. To remove slag

9. What is the best rod combination for repairing a cracked bracket?

A. 6013 root + 6013 cap

B. 6011 root + 7018 cap

C. 7018 only

D. Flux-core only

10. Why are field repairs often reinforced?

A. To make the weld prettier

B. To prevent repeat failures under heavy use

C. To reduce grinding time

D. To save shielding gas

Answer Key

1. B — Wind blows away shielding gas.
2. C — Stick welding is reliable in wind.
3. B — Engine-driven welders provide high amperage anywhere.
4. B — Grind out and bevel the crack before welding.
5. B — Fish plates reinforce cracks.
6. B — Preheating reduces cracking in thick or cold metal.
7. B — Angle grinders are essential for field prep.
8. B — Gussets strengthen corners and brackets.
9. B — 6011 root + 7018 cap is a strong repair combo.
10. B — Reinforcement prevents repeat failures.

CHAPTER 12 — PROFESSIONAL WELDING SKILLS

What You Will Learn

Professional welding goes far beyond laying beads. It requires the ability to read blueprints, interpret welding symbols, understand tolerances, follow WPS documents, and work within inspection standards. These skills are essential in fabrication shops, structural welding, manufacturing, and any environment where welds must meet code.

By the end of this chapter, you will understand:

- How to read and interpret welding blueprints
- How welding symbols communicate joint type, size, and process
- How tolerances affect fit-up and final dimensions
- What WPS (Welding Procedure Specifications) are and how to follow them
- How welds are inspected and evaluated
- The basics of NDT (Non-Destructive Testing) methods

These skills elevate your work from "good" to "professional-grade."

Blueprint Reading

Blueprints are the universal language of fabrication. They tell you what to build, how big it must be, and how the parts fit together.

Key Elements of a Blueprint

- Views: Top, front, side, isometric
- Dimensions: Length, width, height, hole spacing
- Notes: Special instructions
- Material callouts: Type, thickness, grade
- Weld symbols: Type, size, length, and location

Reading Dimensions

- Always read from left to right
- Pay attention to decimal vs. fractional measurements
- Look for reference dimensions (REF)
- Check for tolerances (+/–)

Blueprint reading is a skill that grows with experience.

Welding Symbols

Welding symbols communicate weld requirements without long written instructions.

Basic Components

- Reference line — the main horizontal line
- Arrow — points to the joint
- Tail — optional; used for notes or process info
- Symbol — fillet, groove, plug, spot, etc.
- Numbers — size, length, pitch, angle

Common Weld Symbols

- Fillet weld — triangle
- Groove welds — V, U, J, bevel, square
- Plug/slot welds
- Spot welds
- Seam welds

Arrow Side vs. Other Side

- Arrow side welds go below the reference line
- Other side welds go above the reference line
- Both sides use symbols on both sides

Additional Indicators

- Contour symbols: Flat, convex, concave
- Finish symbols: Grind, machine, chip
- Intermittent welds: Shown as “2–6” (2" weld every 6")

Once you understand welding symbols, you can work anywhere.

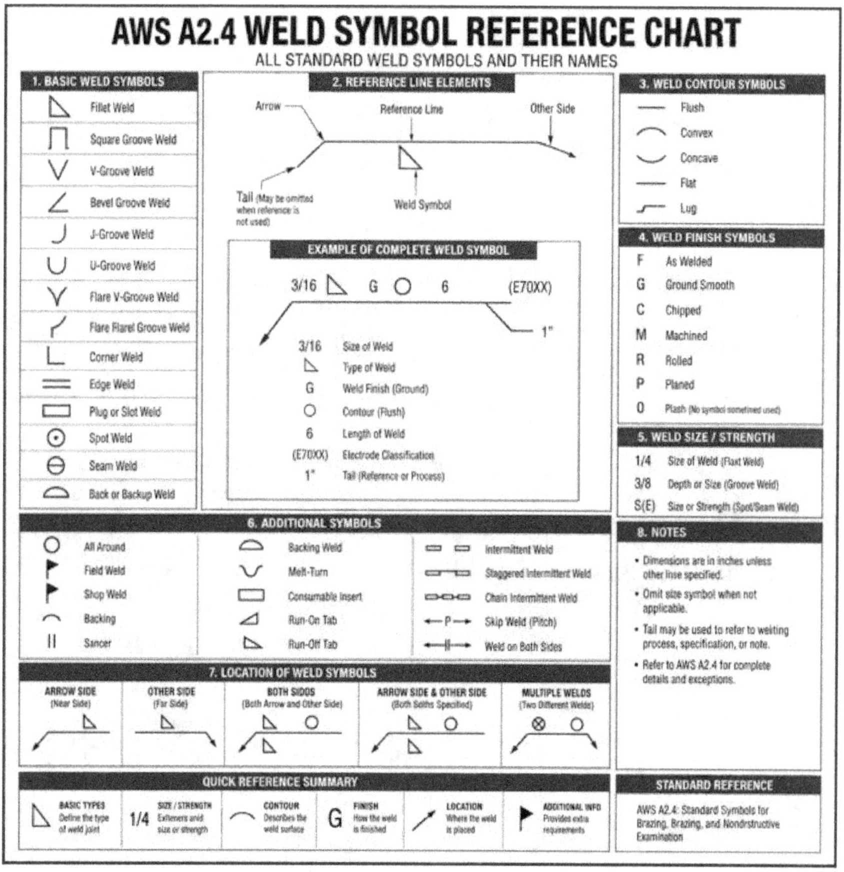

Tolerances

Tolerances define how much variation is allowed in a part's dimensions.

Examples

- ± 1/16"
- ± 0.010"
- ± 1° angle tolerance

Why Tolerances Matter

- Parts must fit together
- Frames must be square
- Holes must align
- Weld shrinkage must be accounted for

Professional welders work *within tolerance*, not "close enough."

WPS Documents (Welding Procedure Specifications)

A WPS is a formal document that tells the welder exactly how to perform a weld.

A WPS Includes:

- Welding process (MIG, TIG, Stick)
- Amperage and voltage range
- Wire or rod type
- Shielding gas
- Joint design
- Preheat and interpass temperature
- Travel speed
- Welding position

Why WPS Matters

- Ensures consistency
- Meets code requirements
- Provides traceability
- Required for structural and certified welding

Following a WPS is a core professional skill.

Inspection

Welds are inspected to ensure they meet quality standards.

Types of Inspection

- Visual Inspection (VT) — first and most important
- Dimensional Inspection — checking size, length, and tolerances
- Weld profile inspection — reinforcement, undercut, etc.

What Inspectors Look For

- Cracks
- Porosity
- Undercut
- Lack of fusion
- Correct size and length
- Proper contour

Good welders inspect their own work before anyone else does.

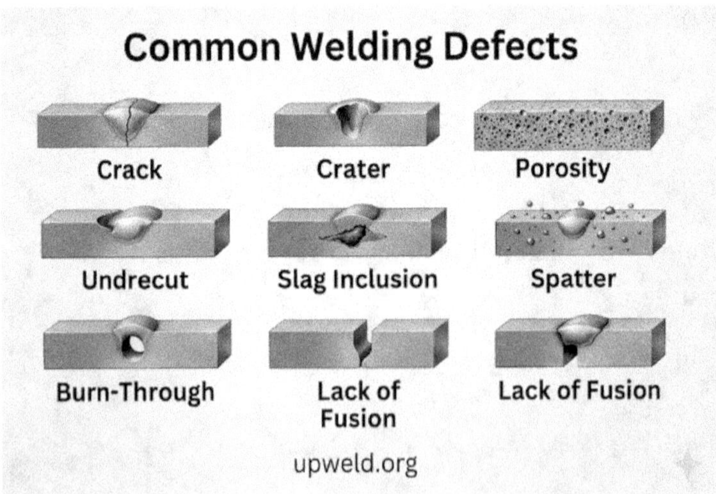

NDT Methods (Non-Destructive Testing)

NDT checks weld quality without damaging the part.

Common NDT Methods

- VT — Visual Testing
- PT — Dye Penetrant Testing
- MT — Magnetic Particle Testing
- UT — Ultrasonic Testing
- RT — Radiographic Testing (X-ray)

What NDT Detects

- Internal cracks
- Lack of fusion
- Porosity
- Incomplete penetration
- Subsurface defects

NDT is essential in structural, pipeline, aerospace, and pressure-vessel welding.

Why Professional Welding Skills Matter

These skills separate hobby welders from professionals. When you can read blueprints, interpret symbols, follow WPS documents, and understand inspection requirements, you can:

- Work in fabrication shops
- Build structural components
- Pass certification tests
- Produce consistent, code-compliant welds
- Advance your welding career

Professional welding is about precision, consistency, and documentation — not just technique.

Chapter 12 — Knowledge Check

Chapter 12 Quiz

1. What does a blueprint primarily communicate?
A. Welding machine settings
B. Dimensions, materials, and how parts fit together
C. Gas flow requirements
D. Electrode storage instructions

2. Where is an arrow-side weld symbol placed?
A. Above the reference line
B. Below the reference line
C. On the tail
D. On the right side only

3. What does a fillet weld symbol look like?
A. A circle
B. A triangle
C. A square
D. A curved line

4. What does a tolerance of ± 1/16" mean?
A. The part must be exact
B. The part can vary by 1/16" in either direction
C. The part can be any size
D. The part must be oversized

5. What is the purpose of a WPS?

A. To choose paint colors

B. To tell the welder exactly how to perform a weld

C. To store electrodes

D. To calculate gas usage

6. Which inspection method is always performed first?

A. Radiographic testing

B. Ultrasonic testing

C. Visual inspection

D. Magnetic particle testing

7. What does MT (Magnetic Particle Testing) detect?

A. Gas flow

B. Surface and near-surface cracks

C. Weld bead height

D. Paint thickness

8. What does the tail of a welding symbol typically contain?

A. Material thickness

B. Notes or process information

C. Weld length

D. Tolerance values

9. What does UT (Ultrasonic Testing) check for?

A. Internal defects

B. Weld color

C. Gas flow

D. Electrode angle

10. Why are tolerances important?

A. They make the blueprint look professional

B. They ensure parts fit and function correctly

C. They reduce welding time

D. They eliminate the need for measuring

Answer Key

1. B
2. B
3. B
4. B
5. B
6. C
7. B
8. B
9. A
10. B

CHAPTER 13 — TROUBLESHOOTING GUIDE

Diagnose Problems. Fix Them Fast. Weld With Confidence.

What You Will Learn

Every welder — beginner or expert — runs into problems. The difference between frustration and mastery is knowing why a defect happens and how to correct it. This chapter gives you a practical, no-nonsense system for identifying welding defects, understanding their causes, and applying the right fix.

By the end of this chapter, you will be able to:

- Recognize the most common welding defects
- Understand the root causes behind each problem
- Adjust heat, travel speed, angle, and technique to correct issues
- Prevent defects before they start
- Troubleshoot MIG, Stick, TIG, and Flux-Core with confidence

This is the chapter welders return to again and again.

Porosity

Tiny holes in or on the weld bead.

What It Looks Like

- Pinholes on the surface
- Wormholes inside the weld
- Weak, brittle welds

Common Causes

- Contaminated metal (rust, oil, paint, moisture)
- Poor shielding gas coverage
- Wind blowing gas away
- Gas flow too low or too high
- Dirty filler rod (TIG)
- Damp electrodes (Stick)

How to Fix It

- Clean metal thoroughly
- Increase gas flow (20–25 CFH for MIG/TIG)
- Block wind or switch to Stick/Flux-Core
- Check hoses, fittings, and regulator for leaks
- Keep tungsten and filler rod clean
- Store rods properly (especially 7018)

Rule: Porosity always comes from contamination or bad shielding.

Spatter

Molten droplets sticking around the weld.

What It Looks Like

- BB-like dots around the bead
- Rough, dirty appearance

Common Causes

- Voltage too low
- Wire speed too high
- Long stickout
- Dirty metal
- Wrong polarity
- Poor technique

How to Fix It

- Increase voltage
- Reduce wire speed
- Shorten stickout (⅜"–½")
- Clean metal
- Verify polarity (MIG = DCEP)

Rule: Spatter is almost always a settings issue.

Undercut

A groove melted into the base metal along the weld toe.

What It Looks Like

- A sharp valley beside the weld
- Weakens the joint significantly

Common Causes

- Too much heat
- Traveling too fast
- Wrong torch/electrode angle
- Poor puddle control

How to Fix It

- Reduce amperage/voltage
- Slow down slightly
- Adjust angle (10–15° push for MIG; drag for Stick/Flux-Core)
- Use a small weave if needed

Rule: Undercut = too hot or too fast.

Lack of Fusion

The weld metal does not bond to the base metal.

What It Looks Like

- Cold, lumpy bead
- Edges not tied in
- Weld sits "on top" of the metal

Common Causes

- Heat too low
- Traveling too fast
- Incorrect angle
- Dirty metal
- Weaving too wide

How to Fix It

- Increase heat
- Slow down
- Improve angle
- Clean metal thoroughly
- Use stringer beads instead of wide weaves

Rule: Lack of fusion is one of the most dangerous defects.

Burn-Through

A hole melted through the metal.

What It Looks Like

- Holes in the joint
- Edges collapsing
- Excessive warping

Common Causes

- Too much heat
- Staying in one spot too long
- Thin metal
- Poor fit-up

How to Fix It

- Lower amperage/voltage
- Increase travel speed
- Use smaller wire (.023" for thin metal)
- Use backing bars or copper plates
- Improve fit-up

Rule: Thin metal requires finesse, not force.

Cracking

Cracks in or near the weld — immediate or delayed.

What It Looks Like

- Straight or jagged cracks
- Appears in the weld or HAZ
- Often shows up after cooling

Common Causes

- High heat input
- Rapid cooling
- Dirty or contaminated metal
- High-carbon steel
- Poor joint design
- Welding over existing cracks

How to Fix It

- Preheat thick or high-carbon steel
- Reduce heat input
- Grind out cracks completely
- Improve joint design
- Use proper filler metal

Rule: Never weld over a crack — remove it first.

Slag Inclusions

Slag trapped inside the weld.

What It Looks Like

- Dark lines or pockets inside the bead
- Weak welds
- Rough appearance

Common Causes

- Wrong angle
- Traveling too fast
- Not cleaning between passes
- Wide weave patterns
- Low heat

How to Fix It

- Maintain correct drag angle (Stick/Flux-Core)
- Slow down
- Clean thoroughly between passes
- Use stringers instead of wide weaves
- Increase heat slightly

Rule: Slag must be removed completely before welding over it.

Summary

Troubleshooting is a core welding skill. When you understand what causes defects — and how to fix them — you gain full control over your welds. Every welder makes mistakes; skilled welders know how to correct them.

This chapter is your roadmap to diagnosing problems quickly and welding with confidence.

Chapter 13 — Knowledge Check

1. What is the most common cause of porosity?

A. Too much amperage

B. Contaminated metal or poor gas coverage

C. Using a large cup

D. Traveling too slowly

2. What causes excessive spatter in MIG welding?

A. Voltage too low

B. Voltage too high

C. Gas flow too high

D. Using a small wire diameter

3. What is undercut?

A. A hole in the weld

B. A groove melted into the base metal along the weld toe

C. Excessive buildup

D. A type of weave pattern

4. What is the main cause of lack of fusion?

A. Too much heat

B. Traveling too fast or heat too low

C. Using too large of a cup

D. Too much gas flow

5. What causes burn-through?
A. Too little heat
B. Too much heat or slow travel
C. Dirty metal
D. Wrong polarity

6. What should you do before repairing a crack?
A. Weld directly over it
B. Grind it out completely
C. Cool it with water
D. Increase gas flow

7. What causes slag inclusions?
A. Too much gas flow
B. Not cleaning between passes or wrong angle
C. Using a small electrode
D. Welding too slowly

8. What is the best way to prevent porosity outdoors?
A. Increase wire speed
B. Block wind or switch to Stick/Flux-Core
C. Use a larger cup
D. Weld faster

9. What is a common fix for undercut?

A. Increase heat

B. Reduce heat and slow down

C. Use a larger wire

D. Switch to TIG

10. What is the safest way to fix a crack in steel?

A. Weld over it

B. Grind it out and re-weld

C. Add more filler metal

D. Increase travel speed

Answer Key

1. B
2. A
3. B
4. B
5. B
6. B
7. B
8. B
9. B
10. B

CHAPTER 14 — MAINTENANCE & CONSUMABLES

What You Will Learn

A welding machine is only as good as the care it receives. Dirty liners, worn contact tips, clogged nozzles, and neglected gas systems cause more welding problems than technique errors. This chapter teaches you how to maintain your equipment, replace consumables, and keep your machine running at peak performance.

By the end of this chapter, you will understand:

- How to maintain MIG, TIG, and Stick machines
- How to inspect and replace contact tips, nozzles, rollers, and liners
- How to care for tungsten electrodes
- How to maintain gas systems and hoses
- How to store consumables properly
- How to prevent common machine-related welding issues

Good maintenance saves time, reduces frustration, and extends the life of your equipment.

Machine Care

Your welder is a tool — and tools need upkeep.

General Maintenance

- Blow out dust with compressed air
- Keep vents clear
- Inspect cables for cuts or burns
- Tighten loose connections
- Check ground clamp for corrosion

MIG Machine Care

- Clean drive rollers
- Replace worn liners
- Keep wire dry and rust-free
- Inspect gun for heat damage

TIG Machine Care

- Keep torch parts clean
- Inspect hoses for leaks
- Replace worn collets and cups
- Maintain pedal and cable connections

Stick Welder Care

- Keep leads clean and flexible
- Inspect electrode holder jaws
- Check for overheating

A clean machine welds better — and lasts longer.

Contact Tips

Contact tips transfer electrical current to the wire. When they wear out, your arc becomes unstable.

Signs a Contact Tip Needs Replaced

- Erratic arc
- Wire stubbing
- Excessive spatter
- Wire feeding inconsistently

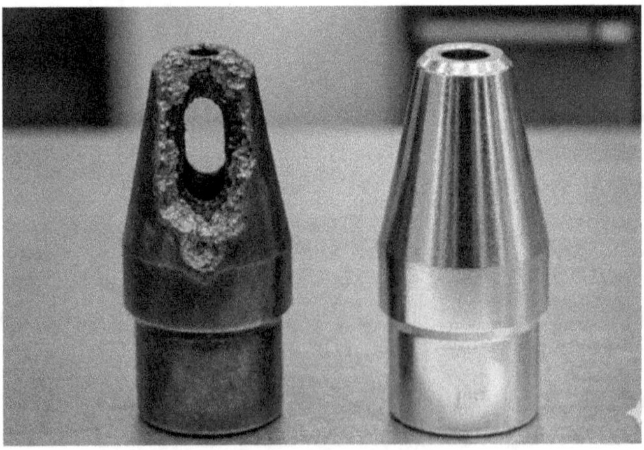

Maintenance

- Replace tips regularly
- Match tip size to wire size
- Keep spare tips on hand

A fresh contact tip often fixes half of a welder's problems.

Nozzles

Nozzles direct shielding gas around the weld.

Common Issues

- Spatter buildup
- Blocked gas flow
- Overheating

Maintenance

- Clean with a nozzle brush
- Use anti-spatter spray
- Replace if cracked or warped

Good gas coverage depends on a clean nozzle.

Drive Rollers

Drive rollers feed wire smoothly through the gun.

Signs of Roller Problems

- Bird-nesting
- Slipping wire
- Inconsistent feed speed

Maintenance

- Clean rollers
- Replace worn rollers
- Match roller type to wire (V-groove, U-groove, knurled)
- Adjust tension properly

Correct roller tension prevents feeding headaches.

Liners

The liner guides wire from the feeder to the gun. A dirty or kinked liner causes feeding issues.

Signs of a Bad Liner

- Wire jams
- Inconsistent arc
- Excessive drag
- Bird-nesting at the feeder

Maintenance

- Replace liners regularly
- Cut liners to correct length
- Keep wire clean to extend liner life

A clean liner equals smooth wire feeding.

Tungsten Electrodes

Tungsten must be clean and properly shaped for TIG welding.

Maintenance

- Grind tungsten on a dedicated wheel
- Grind lengthwise, not sideways
- Keep tungsten free of contamination
- Store in a clean container

Signs Tungsten Needs Re-Grinding

- Arc wandering
- Contaminated tip
- Poor penetration
- Dirty weld puddle

Sharp tungsten = precise TIG control.

Gas Systems

Shielding gas is critical for MIG and TIG.

Maintenance

- Check hoses for leaks
- Inspect regulator gauges
- Keep cylinders upright and secured
- Replace cracked O-rings
- Purge lines before welding

Common Gas Problems

- Porosity
- Dirty welds
- Arc instability

Most gas issues come from leaks or blocked nozzles.

Storage

Proper storage protects consumables and extends their life.

Store Electrodes

- Keep 7018 rods in a rod oven
- Keep 6010/6011 dry
- Seal opened boxes

Store Wire

- Keep spools dry
- Cover when not in use
- Avoid humidity

Store Tungsten

- Use labeled tubes
- Keep clean and separated

Store Tools

- Keep grinders, brushes, and squares clean
- Replace worn discs and wheels

Good storage prevents contamination and saves money.

Why Maintenance Matters

Most welding problems come from equipment issues — not technique.
Proper maintenance:

- Improves weld quality
- Reduces downtime
- Extends machine life
- Prevents frustration
- Saves money on repairs

A well-maintained machine makes welding easier, cleaner, and more enjoyable.

Chapter 14 — Knowledge Check

Chapter 14 Quiz

1. What is the main function of a contact tip?
A. To cool the weld
B. To transfer electrical current to the wire
C. To hold the nozzle in place
D. To regulate gas flow

2. What is a common sign of a worn contact tip?
A. Smooth arc
B. Wire feeding inconsistently
C. Lower gas usage
D. Cleaner welds

3. What causes bird-nesting in a MIG welder?
A. Too much gas flow
B. Worn or dirty drive rollers
C. Using a large nozzle
D. Low amperage

4. What is the purpose of a liner?
A. To hold the tungsten
B. To guide wire from feeder to gun
C. To cool the torch
D. To store electrodes

5. How should tungsten be ground?
A. Sideways
B. Lengthwise
C. In circles
D. With a file

6. What is a sign of a contaminated tungsten?
A. Smooth arc
B. Arc wandering
C. Lower amperage
D. Cleaner puddle

7. What helps prevent porosity caused by gas issues?
A. Using more wire
B. Checking hoses and regulators for leaks
C. Increasing travel speed
D. Using a larger nozzle

8. Why should 7018 rods be stored in a rod oven?
A. To keep them shiny
B. To prevent moisture absorption
C. To increase amperage
D. To reduce spatter

9. What does a clogged nozzle affect?

A. Wire speed

B. Gas coverage

C. Voltage

D. Polarity

10. Why is machine maintenance important?

A. It makes the machine heavier

B. It improves weld quality and reduces problems

C. It increases gas usage

D. It eliminates the need for consumables

Answer Key

1. B
2. B
3. B
4. B
5. B
6. B
7. B
8. B
9. B
10. B

CHAPTER 15 — FINAL THOUGHTS

What You Will Learn

Welding is a lifelong skill — one that grows with every project, every mistake, and every hour behind the hood. This final chapter gives you guidance on how to continue improving, how to choose meaningful projects, how to stay safe, and how to build long-term welding confidence.

By the end of this chapter, you will understand:

- Why consistent practice matters
- How to choose projects that build real skill
- How to maintain a safety-first mindset
- How to track your progress and set goals
- How to continue developing as a welder

This chapter is your roadmap for long-term growth.

Practice

Welding is a hands-on craft. The more you weld, the better you get.

Why Practice Matters

- Builds muscle memory
- Improves puddle control
- Sharpens hand-eye coordination
- Teaches heat management
- Reduces mistakes

How to Practice Effectively

- Practice short sessions often
- Weld in all positions
- Try different metals and thicknesses
- Repeat the same joint until it becomes natural
- Keep scrap metal for practice

Skill comes from repetition — not rushing.

Project Ideas

Projects are the best way to apply what you've learned.

Beginner Projects

- Welding cart
- Small table
- Fire pit
- Tool rack
- Simple brackets

Intermediate Projects

- Workbench
- Trailer repairs
- Gates and fences
- Shelving units
- Shop fixtures

Advanced Projects

- Trailers
- Structural frames
- Custom equipment
- Automotive fabrication
- Heavy repairs

Choose projects that challenge you without overwhelming you.

Skill Growth

Welding is a journey — not a destination.

Ways to Grow Your Skills

- Try new processes (TIG, Stick, Flux-Core)
- Learn to read blueprints
- Practice out-of-position welding
- Study welding symbols
- Take certification tests
- Learn fabrication and layout
- Watch experienced welders

Track Your Progress

- Keep a welding journal
- Record settings and results
- Save samples of your welds
- Note what worked and what didn't

Growth comes from curiosity and consistency.

Safety Mindset

Safety is not optional — it's a habit.

Core Safety Principles

- Wear proper PPE
- Keep your workspace clean
- Ventilate fumes
- Inspect equipment regularly
- Never weld near flammable materials
- Protect your eyes and skin
- Respect electricity and heat

A safe welder is a long-lasting welder.

Long-Term Development

Welding can become a career, a trade, or a lifelong hobby.

Long-Term Goals to Consider

- Earning certifications
- Starting a welding business
- Learning advanced TIG techniques
- Building custom fabrication projects
- Teaching others
- Expanding into machining or metal art

Stay Motivated

- Set achievable goals
- Celebrate progress
- Keep learning
- Stay patient
- Enjoy the process

Welding rewards those who stick with it.

Final Message

You now have the knowledge, techniques, and mindset to weld confidently and safely. Whether you're repairing equipment, building projects, or pursuing welding professionally, remember:

Every great welder started exactly where you are — with a spark, a hood, and a willingness to learn.

Keep practicing.
Keep building.
Keep improving.

Your welding journey has just begun.

Chapter 15 — Knowledge Check

Chapter 15 Quiz

1. What is the most important factor in long-term welding improvement?
A. Expensive equipment
B. Consistent practice
C. Using only one welding process
D. Avoiding mistakes

2. Why are projects important for skill development?
A. They use up scrap metal
B. They apply skills in real situations
C. They eliminate the need for practice
D. They reduce machine maintenance

3. What is one benefit of keeping a welding journal?
A. It replaces the need for training
B. It helps track settings and progress
C. It reduces gas usage
D. It prevents spatter

4. Which project is appropriate for a beginner?
A. Trailer frame
B. Welding cart
C. Automotive roll cage
D. Structural beam

5. Why is a safety mindset important?
A. It makes welds shinier
B. It prevents injuries and accidents
C. It increases amperage
D. It reduces wire usage

6. What is a good way to stay motivated?
A. Avoid challenging projects
B. Compare yourself to professionals
C. Set achievable goals
D. Weld only once a month

7. What is one sign of long-term welding growth?
A. Needing more filler metal
B. Using fewer clamps
C. Improved consistency and control
D. Faster travel speed only

8. What should you do when learning a new welding process?

A. Expect perfection immediately

B. Practice short sessions often

C. Avoid reading instructions

D. Use maximum heat

9. Why is it helpful to save samples of your welds?

A. To recycle them later

B. To compare improvement over time

C. To use them as scrap

D. To reduce grinding

10. What is the best mindset for long-term welding success?

A. Perfection or nothing

B. Slow, steady improvement

C. Only weld when necessary

D. Avoid difficult joints

APPENDIX — QUICK REFERENCE GUIDE

This appendix provides fast access to essential charts, tables, and information welders need on the job. It is designed for quick lookup without flipping through the entire book.

1. MIG Wire Selection Chart

Metal Type	Thickness	Recommended Wire	Polarity	Gas
Mild Steel	22ga–1/8"	.023"–.030" ER70S-6	DCEP	75/25
Mild Steel	1/8"–1/2"	.030"–.035" ER70S-6	DCEP	75/25
Stainless	Thin–Medium	.030" ER308L	DCEP	98/2 or Tri-Mix
Aluminum	Thin	.030"–.035" ER4043	DCEP	100% Argon
Flux-Core	All	.030"–.035" E71T-11	DCEN	None

2. Stick Electrode Quick Chart

Rod	Polarity	Penetration	Best Use
6010	DCEP	Deep	Dirty metal, root passes
6011	AC/DCEP	Deep	Field repairs, rust, paint
6013	AC/DCEP/DCEN	Medium	Sheet metal, smooth beads
7014	AC/DCEP	Medium	Fast fill, production
7018	AC/DCEP	Medium	Structural, clean metal
7024	AC/DCEP	Low	Flat/horizontal heavy fill

3. TIG Tungsten Selection

Tungsten Type	Color	Best For
2% Thoriated	Red	Steel, stainless
2% Ceriated	Gray	Low-amp work, thin metal
2% Lanthanated	Blue	All-purpose, AC/DC
Pure	Green	Aluminum (older machines)

4. Joint Preparation Guide

Joint Type	When to Use	Prep Needed
Butt Joint	Plate, tubing	Bevel >1/8"
Lap Joint	Sheet metal	Clean overlap
T-Joint	Frames, brackets	Clean edges
Corner Joint	Boxes, frames	Fit tight
Edge Joint	Thin metal	Low heat

5. Troubleshooting Quick Fix Table

Problem	Likely Cause	Quick Fix
Porosity	Gas issue	Increase flow, block wind
Spatter	Low voltage	Raise voltage
Undercut	Too hot/fast	Reduce heat, slow down
Burn-Through	Too hot	Lower heat, move faster
Lack of Fusion	Too cold	Increase heat
Slag Inclusion	Dirty or wrong angle	Clean between passes

6. Gas Flow Reference

Process	Gas	Flow Rate
MIG Steel	75/25	20–25 CFH
MIG Stainless	Tri-Mix	25–30 CFH
MIG Aluminum	Argon	25–30 CFH
TIG Steel	Argon	15–20 CFH
TIG Aluminum	Argon	15–20 CFH

7. Preheat Temperature Guide

Metal	Thickness	Preheat
Mild Steel	>1/2"	150–250°F
High-Carbon Steel	Any	300–500°F
Cast Iron	Any	500–1200°F
Aluminum	Thick	200–300°F

8. Conversion & Measurement Table

Fractions to Decimals

- 1/16" = 0.0625
- 1/8" = 0.125
- 3/16" = 0.1875
- 1/4" = 0.250
- 3/8" = 0.375
- 1/2" = 0.500

Common Angles

- 15° = Light bevel
- 30° = Standard bevel
- 45° = Heavy bevel

9. Safety Checklist

Before Welding

- PPE on
- Area clear of flammables
- Ventilation running
- Ground clamp secure
- Machine settings correct

After Welding

- Check for smoldering material
- Coil leads safely
- Turn off gas
- Power down machine

10. Field Repair Kit Checklist

- 6011 rods
- 7018 rods
- Grinder + discs
- Wire brush
- Chipping hammer
- Clamps
- Soapstone
- Tape measure
- Portable welder or generator

www.ingramcontent.com/pod-product-compliance
Lightning Source LLC
LaVergne TN
LVHW022212200926
848481LV00006B/1939
9798996646906